© 2017
Clement Ampadu
drampadu@hotmail.com

ISBN: 978-1-387-06611-7
ID: 21108813
www.lulu.com

All rights reserved. No part of this publication may be produced or transmitted in any form or by any means, electronic or mechanical, including photocopying and recording, or in any information storage and retrieval system, without the prior written permission of the publisher.

Contents

Preface   3

Dedication   4

1 Fixed Point Theorems for Implicit Chatterjea-type Cyclic Weakly Multiplicative Contraction in Multiplicative Metric Spaces   5
1.1 Brief Summary 5
1.2 Preliminaries 5
1.3 Main Results 6
1.4 Exercises 12
1.5 References 13

2 Fixed Point Results for Implicit Cyclic Weak φ-Multiplicative Contractions on Multiplicative Partial Metric Spaces   14
2.1 Brief Summary 14
2.2 Preliminaries 14
2.3 Main Results 16
2.4 Exercises 21
2.5 References 22

3 Existence and Uniqueness of Fixed Points for Implicit Almost Generalized Cyclic (ψ, φ)-Weak Multiplicative Contractive Type Mappings   24
3.1 Brief Summary 24
3.2 Preliminaries 24
3.3 Main Results 25
3.4 Exercises 29
3.5 References 31

4 Fixed Point Theorems for Implicit Cyclic Weak φ_G-Multiplicative Contraction Mappings   33
4.1 Brief Summary 33
4.2 Preliminaries 33
4.3 Main Results 34
4.4 Exercises 38
4.5 References 40

Preface

The notion of C-class function was introduced by A.H. Ansari [A.H. Ansari, Note on $\varphi - \psi$-contractive type mappings and related fixed point, The 2nd Regional Conference on Mathematics and Applications, PNU, September 2014, 377-380]. On the other hand, the notion of multiplicative C-class function was initiated by C.B. Ampadu and then jointly by Ampadu and Ansari [Clement Ampadu and Arslan Hojat Ansari, FIXED POINT THEOREMS IN COMPLETE MULTIPLICATIVE METRIC SPACES WITH APPLICATION TO MULTIPLICATIVE ANALOGUE OF C-CLASS FUNCTIONS, JP Journal of Fixed Point Theory and Applications, August 2016, Volume 11, Issue 2, Pages 113 - 124].

Fixed point theorems for mappings satisfying cyclical contractive conditions first appeared in [W.A. Kirk, P.S. Srinivasan, P. Veeramani, Fixed points for mappings satisfying cyclical contractive conditions, Fixed Point Theory. 4 (2003) 79-89], and in this monograph we have defined the multiplicative version of cyclic extension of weakly contractive mappings [Ya. I. Alber and S. Guerr-Delabriere, Principle of weakly contractive maps Hilbert spaces, New Results in Operator Theory and its Applications (I.Gohberg and Yu. Lyubich, eds.), Oper. Theory Adv. Appl., vol. 98, Birkhauser, Basel, 1997, pp. 7–22] implicitly via the multiplicative C-class function of Ampadu and Ansari [Clement Ampadu and Arslan Hojat Ansari, FIXED POINT THEOREMS IN COMPLETE MULTIPLICATIVE METRIC SPACES WITH APPLICATION TO MULTIPLICATIVE ANALOGUE OF C-CLASS FUNCTIONS, JP Journal of Fixed Point Theory and Applications, August 2016, Volume 11, Issue 2, Pages 113 - 124], and obtained some fixed point results in the multiplicative analogue of Metric Spaces (Chapter 1, Chapter 3, and Chapter 4) and Partial Metric Spaces (Chapter 2)

A nice feature of this monograph are the (publishable) exercise set, which begs the reader to explore the beautiful connection between the cyclic extension of weakly contractive mappings, c-class function, and their multiplicative analogue. The reader will find the notion of multiplicative metric space [A.E. Bashirov, E.M. Kurpnar and A. Ozyapc, Multiplicative calculus and its applications J. Math. Anal. Appl., 337 (2008), 36-48] useful as he or she begins his or her own investigative inquiry.

Clement Kwasi Boateng Ampadu

Dedication

Thanking Yahweh, I dedicate this monograph to those who read it, including family, friends, and love ones .

Clement Kwasi Boateng Ampadu
July, 2017

Chapter 1

Fixed Point Theorems for Implicit Chatterjea-type Cyclic Weakly Multiplicative Contraction in Multiplicative Metric Spaces

1.1 Brief Summary

> **Abstract A.1 1**
>
> In this chapter we define the multiplicative version of Chatterjea-type cyclic weakly contraction implicitly via the C-class function of Ampadu and Ansari [Clement Ampadu and Arslan Hojat Ansari, FIXED POINT THEOREMS IN COMPLETE MULTIPLICATIVE METRIC SPACES WITH APPLICATION TO MULTIPLICATIVE ANALOGUE OF C-CLASS FUNCTIONS, JP Journal of Fixed Point Theory and Applications, August 2016, Volume 11, Issue 2, Pages 113-124]. Consequently we obtain some existence and uniqueness results for such mappings in the setting of multiplicative metric spaces.

1.2 Preliminaries

> **Definition A.1 1**
>
> [Kirk, WA, Srinivasan, PS, Veeramani, P: Fixed points for mappings satisfying cyclical contractive conditions. Fixed Point Theory Appl. 4(1), 79-89 (2003)] Let X be a nonempty set and $T: X \mapsto X$ be an operator. We say $X = \bigcup_{i=1}^{m} X_i$ is a cyclic representation of X with respect to T if
>
> (a) X_i; $i = 1, \cdots, m$ are nonempty sets
>
> (b) $T(X_1) \subset X_2, \cdots, T(X_{m-1}) \subset X_m, T(X_m) \subset X_1$

> **Notation A.2 1**
>
> Ψ will denote the class of all monotone increasing continuous functions $\psi: [1, \infty) \mapsto [1, \infty)$ with $\psi(t) = 1$ iff $t = 1$

> **Notation A.3 1**
>
> Φ will denote the class of all lower semi-continuous functions $\phi: [1, \infty)^2 \mapsto [1, \infty)$ with $\phi(a, b) > 1$ for $a, b \in (1, \infty)$ and $\phi(1, 1) = 1$

CHAPTER 1. FIXED POINT THEOREMS FOR IMPLICIT CHATTERJEA-TYPE CYCLIC WEAKLY MULTIPLICATIVE CONTRACTION IN MULTIPLICATIVE METRIC SPACES

Definition A.4 1

Let (X, m) be a multiplicative metric space, k be a natural number, $A_1, \cdots, A_k$ be nonempty subsets of X and $Y = \bigcup_{i=1}^{k} A_i$. An operator $T : Y \mapsto Y$ will be called an implicit Chatterjea-type cyclic weakly multiplicative contraction if

(a) $\bigcup_{i=1}^{k} A_i$ is a cyclic representation of Y with respect to T

(b) $\psi(m(Tx, Ty)) \leq F\left[\dfrac{\psi\left(\sqrt{m(x,Ty) \cdot m(y,Tx)}\right)}{\phi(m(x,Ty), m(y,Tx))}\right]$

for any $x \in A_i$, $y \in A_{i+1}$, $i = 1, \cdots, k$, where $A_{k+1} = A_1$, $\psi \in \Psi$, $\phi \in \Phi$, and $F(x,y) := F\left(\frac{x}{y}\right)$ is a multiplicative C-class function [Clement Ampadu and Arslan Hojat Ansari, FIXED POINT THEOREMS IN COMPLETE MULTIPLICATIVE METRIC SPACES WITH APPLICATION TO MULTIPLICATIVE ANALOGUE OF C-CLASS FUNCTIONS, JP Journal of Fixed Point Theory and Applications, August 2016, Volume 11, Issue 2, Pages 113-124]

1.3 Main Results

Remark A.1 1

$\equiv_*$ will denote what we call multiplicative congruence, and we will say $r - q \equiv_* 1 \pmod{k}$ if $\left(\frac{r}{q+1}\right)^{\frac{1}{k}}$ is an integer. In general we say $a \equiv_* b \pmod{k}$ if $\left(\frac{a}{b}\right)^{\frac{1}{k}}$ is an integer

Lemma A.2 1

For every $\epsilon > 1$, there exists a natural number n such that if $r, q \geq n$ with $r - q \equiv_* 1 \pmod{k}$, then, $m(x_r, x_q) < \epsilon$

> **Proof of Lemma A.2.1**
>
> Suppose not, then there exists $\epsilon > 1$ such that for any $n \in \mathbb{N}$ we can find $r_n > q_n \geq n$ with $r_n - q_n \equiv_* 1 \pmod{k}$ satisfying $m(x_{r_n}, x_{q_n}) \geq \epsilon$. Now, we take $n > k^2$, then corresponding to $q_n \geq n$, we can choose r_n such that it is the smallest integer with $r_n > q_n$ satisfying $r_n - q_n \equiv_* 1 \pmod{k}$ and $m(x_{r_n}, x_{q_n}) \geq \epsilon$. Therefore $m(x_{r_n-k}, x_{q_n}) < \epsilon$. By the multiplicative triangle inequality, we have,
>
> $$\epsilon \leq m(x_{q_n}, x_{r_n})$$
> $$\leq m(x_{q_n}, x_{r_n-k}) \cdot \prod_{i=1}^{k} m(x_{r_n-i}, x_{r_n-i+1})$$
> $$< \epsilon \cdot \prod_{i=1}^{k} m(x_{r_n-i}, x_{r_n-i+1})$$
>
> Now taking limits in the above as $n \to \infty$ and assuming that $m(x_{n+1}, x_n) \to 1$, we obtain $\lim_{n \to \infty} m(x_{q_n}, x_{r_n}) = \epsilon$. Again by the multiplicative triangle inequality, we have,
>
> $$\epsilon \leq m(x_{q_n}, x_{r_n})$$
> $$\leq m(x_{q_n}, x_{q_n+1}) \cdot m(x_{q_n+1}, x_{r_n+1}) \cdot m(x_{r_n+1}, x_{r_n})$$
> $$\leq m(x_{q_n}, x_{q_n+1})^2 \cdot m(x_{q_n}, x_{r_n}) \cdot m(x_{r_n}, x_{r_n+1})^2$$
>
> Now taking limits in the above and assuming that $m(x_{n+1}, x_n) \to 1$, we get
>
> $$\lim_{n \to \infty} m(x_{q_n+1}, x_{r_n+1}) = \epsilon$$
>
> Now observe that
>
> $$m(x_{q_n}, Tx_{r_n}) = m(x_{q_n}, x_{r_n+1})$$
> $$\leq m(x_{q_n}, x_{r_n}) \cdot m(x_{r_n}, x_{r_n+1})$$
>
> and
>
> $$m(x_{r_n}, Tx_{q_n}) = m(x_{r_n}, x_{q_n+1})$$
> $$\leq m(x_{r_n}, x_{q_n}) \cdot m(x_{q_n}, x_{q_n+1})$$
>
> If we take limits in the two inequalities immediately above, we deduce that
>
> $$\lim_{n \to \infty} m(x_{q_n}, Tx_{r_n}) = \epsilon \text{ and } \lim_{n \to \infty} m(x_{r_n}, Tx_{q_n}) = \epsilon$$
>
> Since x_{q_n} and x_{r_n} lie in differently adjacently labeled sets A_i and A_{i+1} for certain $1 \leq i \leq k$, then using the fact that T is an implicit Chatterjea-type cyclic weakly multiplicative contraction, we deduce that
>
> $$\psi(\epsilon) \leq \psi(m(x_{q_n+1}, x_{r_n+1}))$$
> $$= \psi(m(Tx_{q_n}, Tx_{r_n}))$$
> $$\leq F\left[\frac{\psi(\sqrt{m(x_{q_n}, Tx_{r_n}) \cdot m(x_{r_n}, Tx_{q_n})})}{\phi(m(x_{q_n}, Tx_{r_n}), m(x_{r_n}, Tx_{q_n}))}\right]$$
> $$= F\left[\frac{\psi(\sqrt{m(x_{q_n}, x_{r_n+1}) \cdot m(x_{r_n}, x_{q_n+1})})}{\phi(m(x_{q_n}, x_{r_n+1}), m(x_{r_n}, x_{q_n+1}))}\right]$$
>
> Since $\lim_{n \to \infty} m(x_{q_n}, Tx_{r_n}) = \epsilon$ and $\lim_{n \to \infty} m(x_{r_n}, Tx_{q_n}) = \epsilon$, then by the continuity of ψ and lower semi-continuity of ϕ, if we take limits in the above inequality we deduce that $\psi(\epsilon) \leq F\left[\frac{\psi(\sqrt{\epsilon^2})}{\phi(\epsilon,\epsilon)}\right]$ which implies $\phi(\epsilon, \epsilon) = 1$ and hence $\epsilon = 1$ which is contrary to $\epsilon > 1$, and the result follows.

Theorem A.3 1

Let (X, m) be a complete multiplicative metric space, $k \in \mathbb{N}$, $A_1, \cdots, A_k$ be nonempty closed subsets of X and $Y = \bigcup_{i=1}^{k} A_i$. Suppose T is an implicit Chatterjea-type cyclic weakly multiplicative contraction, then T has a unique fixed point $z \in \bigcap_{i=1}^{k} A_i$

Proof of Theorem A.3.1

Let $x_0 \in X$, we can construct a sequence $\{x_n\}$ such that $x_{n+1} = Tx_n$, $n = 0, 1, 2, \cdots$. If there exists $n_0 \in \mathbb{N}$ such that $x_{n_0+1} = x_{n_0}$, then it follows that $Tx_{n_0} = x_{n_0+1} = x_{n_0}$, and we are done. Now we assume that $x_{n+1} \neq x_n$ for any $n = 0, 1, 2, \cdots$. As $X = \bigcup_{i=1}^{k} A_i$, for any $n > 0$, there exists $i_n \in \{1, 2, \cdots, k\}$ such that $x_{n-1} \in A_{i_n}$ and $x_n \in A_{i_{n+1}}$. Since T is an implicit Chatterjea-type cyclic weakly multiplicative contraction, then by the properties of F, we deduce the following

$$\psi(m(x_{n+1}, x_n)) = \psi(m(Tx_n, Tx_{n-1}))$$
$$\leq F\left[\frac{\psi(\sqrt{m(x_n, Tx_{n-1}) \cdot m(x_{n-1}, Tx_n)})}{\phi(m(x_n, Tx_{n-1}), m(x_{n-1}, Tx_n))}\right]$$
$$= F\left[\frac{\psi(\sqrt{m(x_{n-1}, x_{n+1})})}{\phi(1, m(x_{n-1}, x_{n+1}))}\right]$$
$$\leq \psi(\sqrt{m(x_{n-1}, x_{n+1})})$$

Since ψ is a non-decreasing function, for all $n = 1, 2, \cdots$, we have

$$m(x_{n+1}, x_n) \leq \sqrt{m(x_{n-1}, x_{n+1})} \leq \sqrt{m(x_{n-1}, x_n) \cdot m(x_n, x_{n+1})}$$

From the above it follows that $m(x_{n+1}, x_n) \leq m(x_n, x_{n-1})$, thus the sequence $\{m(x_{n+1}, x_n)\}$ is a monotone decreasing sequence of non-negative real numbers and hence is convergent. It follows that there exists $r \geq 1$ such that $\lim_{n \to \infty} m(x_{n+1}, x_n) = r$. If we take limits as $n \to \infty$ in the inequality below

$$m(x_{n+1}, x_n) \leq \sqrt{m(x_{n-1}, x_{n+1})} \leq \sqrt{m(x_{n-1}, x_n) \cdot m(x_n, x_{n+1})}$$

we deduce that $\lim_{n \to \infty} m(x_{n-1}, x_{n+1}) = r^2$. By the continuity of ψ, lower semi-continuity of ϕ, and taking limits in

$$\psi(m(x_{n+1}, x_n)) = \psi(m(Tx_n, Tx_{n-1}))$$
$$\leq F\left[\frac{\psi(\sqrt{m(x_n, Tx_{n-1}) \cdot m(x_{n-1}, Tx_n)})}{\phi(m(x_n, Tx_{n-1}), m(x_{n-1}, Tx_n))}\right]$$
$$= F\left[\frac{\psi(\sqrt{m(x_{n-1}, x_{n+1})})}{\phi(1, m(x_{n-1}, x_{n+1}))}\right]$$
$$\leq \psi(\sqrt{m(x_{n-1}, x_{n+1})})$$

we deduce that $\psi(r) \leq F\left[\frac{\psi(r)}{\phi(1, r^2)}\right]$. From which it follows that $\phi(1, r^2) = 1$, and hence $r = 1$. Thus, we have $\lim_{n \to \infty} m(x_{n+1}, x_n) = 1$. Now we show that $\{x_n\}$ is a multiplicative Cauchy sequence in Y. For this reason, fix $\epsilon > 1$. By Lemma A.2, we can find $n_0 \in \mathbb{N}$ such that $r, q \geq n_0$ with $r - q \equiv_* 1 \pmod{k}$ and $m(x_r, x_q) \leq \sqrt{\epsilon}$. Since $\lim_{n \to \infty} m(x_n, x_{n+1}) = 1$, we can also find $n_1 \in \mathbb{N}$ such that $m(x_n, x_{n+1}) \leq \sqrt[k]{\sqrt{\epsilon}}$ for any $n \geq n_1$. Assume that $r, s \geq \max\{n_0, n_1\}$ and $s > r$, then there exists $v \in \{1, 2, \cdots, k\}$ such that $s - r \equiv_* v \pmod{k}$, and thus, $s - r + (k - v + 1) \equiv_* 1 \pmod{k}$. It follows that

$$m(x_r, x_s) \leq m(x_r, x_{s+j}) \cdot m(x_{s+j}, x_{s+j-1}) \cdot \ldots \cdot m(x_{s+1}, x_s)$$

Since $m(x_r, x_q) \leq \sqrt{\epsilon}$ and $m(x_n, x_{n+1}) \leq \sqrt[k]{\sqrt{\epsilon}}$, then in conjunction with the above inequality, we deduce that

$$m(x_r, x_s) \leq (\sqrt{\epsilon})^{1 + \frac{j}{k}} \leq (\sqrt{\epsilon})^{1 + \frac{k}{k}} = (\sqrt{\epsilon})^2 = \epsilon$$

Proof of Theorem A.3 Continued 1

It follows that $\{x_n\}$ is a multiplicative Cauchy sequence in Y. Since Y is closed in X, then Y is also complete and there exists $x \in Y$ such that $\lim_{n \to \infty} x_n = x$. Now we show that x is a fixed point of T. Since $Y = \bigcup_{i=1}^{k} A_i$ is a cyclic representation of Y with respect to T, it follows that $\{x_n\}$ has infinite terms in each A_i for $i \in \{1, 2, \cdots, k\}$. Suppose that $x \in A_i$ and $Tx \in A_{i+1}$, and we take subsequence $\{x_{n_q}\}$ of $\{x_n\}$ with $x_{n_q} \in A_i$. By the contractive condition of the theorem, we deduce the following

$$\psi(m(x_{n_q+1}, Tx)) = \psi(m(Tx_{n_q}, Tx))$$
$$\leq F\left[\frac{\psi(\sqrt{m(x_{n_q}, Tx) \cdot m(x, Tx_{n_q})})}{\phi(m(x_{n_q}, Tx), m(x, Tx_{n_q}))}\right]$$
$$= F\left[\frac{\psi(\sqrt{m(x_{n_q}, Tx) \cdot m(x, x_{n_q+1})})}{\phi(m(x_{n_q}, Tx), m(x, x_{n_q+1}))}\right]$$

Now using the continuity of ψ and the lower semi-continuity of ϕ, if we take limits in the above as $n \to \infty$, we deduce that $\psi(m(x, Tx)) \leq F\left[\frac{\psi(\sqrt{m(x,Tx)})}{\phi(m(x,Tx),1)}\right]$ which is a contradiction unless $m(x, Tx) = 1$. It follows that x is a fixed point of T. Now we will prove uniqueness of the fixed point. Suppose that x_1 and x_2 are two fixed points of T and $x_1 \neq x_2$. By the continuity of ψ, lower semi-continuity of ϕ, and the properties of F, we deduce from the contractive condition of the theorem that

$$\psi(m(x_1, x_2)) = \psi(m(Tx_1, Tx_2))$$
$$\leq F\left[\frac{\psi(\sqrt{m(x_1, Tx_2) \cdot m(x_2, Tx_1)})}{\phi(m(x_1, Tx_2), m(x_2, Tx_1))}\right]$$
$$= F\left[\frac{\psi(\sqrt{m(x_1, x_2) \cdot m(x_2, x_1)})}{\phi(m(x_1, x_2), m(x_2, x_1))}\right]$$
$$= F\left[\frac{\psi(m(x_1, x_2))}{\phi(m(x_1, x_2), m(x_2, x_1))}\right]$$
$$\leq \psi(m(x_1, x_2))$$

which is a contradiction unless $x_1 = x_2$, and the theorem follows.

If ψ is the identity in the previous Theorem, then we get the following

Corollary A.4 1

Let (X, m) be a complete multiplicative metric space, $k \in \mathbb{N}$, $A_1, A_2, \cdots, A_k$ be non-empty closed subsets of X and $Y = \bigcup_{i=1}^{k} A_i$. Suppose that $T : Y \mapsto Y$ is an operator such that

(a) $\bigcup_{i=1}^{k} A_i$ is a cyclic representation of Y with respect to T

(b) $m(Tx, Ty) \leq F\left[\frac{\sqrt{m(x,Ty) \cdot m(y,Tx)}}{\phi(m(x,Ty), m(y,Tx))}\right]$

for any $x \in A_i$, $y \in A_{i+1}$, $i = 1, \cdots, k$, where $A_{k+1} = A_1$, $\phi \in \Phi$, and $F(x, y) := F\left(\frac{x}{y}\right)$ is a multiplicative C-class function [Clement Ampadu and Arslan Hojat Ansari, FIXED POINT THEOREMS IN COMPLETE MULTIPLICATIVE METRIC SPACES WITH APPLICATION TO MULTIPLICATIVE ANALOGUE OF C-CLASS FUNCTIONS, JP Journal of Fixed Point Theory and Applications, August 2016, Volume 11, Issue 2, Pages 113-124]. Then has a unique fixed point $z \in \bigcap_{i=1}^{k} A_i$

If in the previous Corollary, we take $F(x, y) := F\left(\frac{x}{y}\right) = \frac{x}{y}$ and let $\phi(a, b) = (a \cdot b)^{\frac{1}{2} - \lambda}$, where $\lambda \in [0, \frac{1}{2})$, then we get the following

Corollary A.5 1

Let (X, m) be a complete multiplicative metric space, $k \in \mathbb{N}$, $A_1, A_2, \cdots, A_k$ be non-empty closed subsets of X and $Y = \bigcup_{i=1}^{k} A_i$. Suppose that $T : Y \mapsto Y$ is an operator such that

(a) $\bigcup_{i=1}^{k} A_i$ is a cyclic representation of Y with respect to T

(b) $m(Tx, Ty) \leq (m(x, Ty) \cdot m(y, Tx))^{\lambda}$

for any $x \in A_i$, $y \in A_{i+1}$, $i = 1, \cdots, k$, where $A_{k+1} = A_1$. Then has a unique fixed point $z \in \bigcap_{i=1}^{k} A_i$

Multiplicative integral was introduced in [A. E. Bashirov, E. M. Kurpinar and A. Ozyapici, Multiplicative calculus and its applications, J. Math. Anal. Appl. 337(1) (2008), 36-48] and we gave an equivalent interpretation [Clement Ampadu, Arslan Hojat Ansari and Memudu Olaposi Olatinwo, FIXED POINT THEOREMS USING MULTIPLICATIVE CONTRACTIVE DEFINITIONS WITH APPLICATION TO MULTIPLICATIVE ANALOGUE OF C-CLASS FUNCTIONS, JP Journal of Fixed Point Theory and Applications, Volume 12, Number 1, 2017, Pages 1-35]. Now letting Ξ be the class of all functions $\xi : [1, \infty) \mapsto [1, \infty)$ such that ξ is a Lebesgue integrable mapping which is summable and $\xi \geq 1$ satisfies $\int_{1}^{\epsilon} \xi(t) dt > 1$ for each $\epsilon > 1$, then we have the following integral type representation of the previous Corollary

Corollary A.6 1

Let (X, m) be a complete multiplicative metric space, $k \in \mathbb{N}$, $A_1, A_2, \cdots, A_k$ be non-empty closed subsets of X and $Y = \bigcup_{i=1}^{k} A_i$. Suppose that $T : Y \mapsto Y$ is an operator such that

(a) $\bigcup_{i=1}^{k} A_i$ is a cyclic representation of Y with respect to T

(b) there exist $\lambda \in [0, \frac{1}{2})$ such that

$$\int_{1}^{m(Tx,Ty)} \xi(s) ds \leq \left(\int_{1}^{(m(x,Ty) \cdot m(y,Tx))} \xi(s) ds \right)^{\lambda}$$

for any $x \in A_i$, $y \in A_{i+1}$, $i = 1, \cdots, k$, where $A_{k+1} = A_1$, and $\xi \in \Xi$. Then has a unique fixed point $z \in \bigcap_{i=1}^{k} A_i$

If in the previous Corollary, we take $A_i = X$ for $i = 1, 2, \cdots, k$, then we get the following

Corollary A.7 1

Let (X, m) be a complete multiplicative metric space. Suppose that $T : X \mapsto X$ is an operator such that

$$\int_{1}^{m(Tx,Ty)} \xi(s) ds \leq \left(\int_{1}^{(m(x,Ty) \cdot m(y,Tx))} \xi(s) ds \right)^{\lambda}$$

for any $x, y \in X$, $\lambda \in [0, \frac{1}{2})$. Then has a unique fixed point $z \in X$

1.4 Exercises

Exercise A.1 1

Let (X, m) be a multiplicative metric space, k be a natural number, $A_1, \cdots, A_k$ be nonempty subsets of X and $Y = \bigcup_{i=1}^{k} A_i$. An operator $T : Y \mapsto Y$ will be called an implicit Reich-type cyclic weakly multiplicative contraction if

(a) $\bigcup_{i=1}^{k} A_i$ is a cyclic representation of Y with respect to T

(b) $\psi(m(Tx, Ty)) \leq F\left[\dfrac{\psi\left(\sqrt[3]{m(x,y) \cdot m(x,Tx) \cdot m(y,Ty)}\right)}{\phi(m(x,y), m(x,Tx), m(y,Ty))}\right]$

for any $x \in A_i$, $y \in A_{i+1}$, $i = 1, \cdots, k$, where $A_{k+1} = A_1$, $\psi \in \Psi$, $\phi : [1, \infty)^3 \mapsto [1, \infty)$ is a lower semi-continuous function with $\phi(a, b, c) > 1$ for $a, b, c \in (1, \infty)$ and $\phi(1, 1, 1) = 1$, and $F(x, y) := F(\frac{x}{y})$ is a multiplicative C-class function [Clement Ampadu and Arslan Hojat Ansari, FIXED POINT THEOREMS IN COMPLETE MULTIPLICATIVE METRIC SPACES WITH APPLICATION TO MULTIPLICATIVE ANALOGUE OF C-CLASS FUNCTIONS, JP Journal of Fixed Point Theory and Applications, August 2016, Volume 11, Issue 2, Pages 113-124]. Now

(i) Prove the following: Let (X, m) be a complete multiplicative metric space, $k \in \mathbb{N}$, $A_1, \cdots, A_k$ be nonempty closed subsets of X and $Y = \bigcup_{i=1}^{k} A_i$. Suppose T is an implicit Reich-type cyclic weakly multiplicative contraction, then T has a unique fixed point $z \in \bigcap_{i=1}^{k} A_i$

(ii) State the Corollary arising from (i) if ψ is the identity

(iii) State the Corollary arising from (ii) if we take $F(x, y) := F(\frac{x}{y}) = \frac{x}{y}$ and let $\phi(a, b, c) = (a \cdot b \cdot c)^{\frac{1}{3} - \lambda}$, where $\lambda \in [0, \frac{1}{3})$

(iv) State the Corollary arising from the integral type representation of (iii)

(v) State the Corollary arising from (iv) if $A_i = X$ for $i = 1, 2, \cdots, k$

CHAPTER 1. FIXED POINT THEOREMS FOR IMPLICIT CHATTERJEA-TYPE CYCLIC WEAKLY MULTIPLICATIVE CONTRACTION IN MULTIPLICATIVE METRIC SPACES

> **Exercise A.2 1**
>
> Let (X, m) be a multiplicative metric space, k be a natural number, $A_1, \cdots, A_k$ be nonempty subsets of X and $Y = \bigcup_{i=1}^{k} A_i$. An operator $T : Y \mapsto Y$ will be called an implicit Hardy and Rogers-type cyclic weakly multiplicative contraction if
>
> (a) $\bigcup_{i=1}^{k} A_i$ is a cyclic representation of Y with respect to T
>
> (b) $\psi(m(Tx, Ty)) \leq F\left[\dfrac{\psi\left(\sqrt[5]{m(x,y)\cdot m(x,Tx)\cdot m(y,Ty)\cdot m(x,Ty)\cdot m(y,Tx)}\right)}{\phi(m(x,y),m(x,Tx),m(y,Ty),m(x,Ty),m(y,Tx))}\right]$
>
> for any $x \in A_i$, $y \in A_{i+1}$, $i = 1, \cdots, k$, where $A_{k+1} = A_1$, $\psi \in \Psi$, $\phi : [1, \infty)^5 \mapsto [1, \infty)$ is a lower semi-continuous function with $\phi(a, b, c, d, e) > 1$ for $a, b, c, d, e \in (1, \infty)$ and $\phi(1, 1, 1, 1, 1) = 1$, and $F(x, y) := F\left(\frac{x}{y}\right)$ is a multiplicative C-class function [Clement Ampadu and Arslan Hojat Ansari, FIXED POINT THEOREMS IN COMPLETE MULTIPLICATIVE METRIC SPACES WITH APPLICATION TO MULTIPLICATIVE ANALOGUE OF C-CLASS FUNCTIONS, JP Journal of Fixed Point Theory and Applications, August 2016, Volume 11, Issue 2, Pages 113-124]. Now
>
> (i) Prove the following: Let (X, m) be a complete multiplicative metric space, $k \in \mathbb{N}$, $A_1, \cdots, A_k$ be nonempty closed subsets of X and $Y = \bigcup_{i=1}^{k} A_i$. Suppose T is an implicit Hardy and Rogers-type cyclic weakly multiplicative contraction, then T has a unique fixed point $z \in \bigcap_{i=1}^{k} A_i$
>
> (ii) State the Corollary arising from (i) if ψ is the identity
>
> (iii) State the Corollary arising from (ii) if we take $F(x, y) := F\left(\frac{x}{y}\right) = \frac{x}{y}$ and let $\phi(a, b, c, d, e) = (a \cdot b \cdot c \cdot d \cdot e)^{\frac{1}{5} - \lambda}$, where $\lambda \in [0, \frac{1}{5})$
>
> (iv) State the Corollary arising from the integral type representation of (iii)
>
> (v) State the Corollary arising from (iv) if $A_i = X$ for $i = 1, 2, \cdots, k$

1.5 References

(1) Clement Ampadu and Arslan Hojat Ansari, FIXED POINT THEOREMS IN COMPLETE MULTIPLICATIVE METRIC SPACES WITH APPLICATION TO MULTIPLICATIVE ANALOGUE OF C-CLASS FUNCTIONS, JP Journal of Fixed Point Theory and Applications, August 2016, Volume 11, Issue 2, Pages 113-124

(2) Kirk, WA, Srinivasan, PS, Veeramani, P: Fixed points for mappings satisfying cyclical contractive conditions. Fixed Point Theory Appl.4(1), 79-89 (2003)

(3) A. E. Bashirov, E. M. Kurpinar and A. Ozyapici, Multiplicative calculus and its applications, J. Math. Anal. Appl. 337(1) (2008), 36-48

(4) Clement Ampadu, Arslan Hojat Ansari and Memudu Olaposi Olatinwo, FIXED POINT THEOREMS USING MULTIPLICATIVE CONTRACTIVE DEFINITIONS WITH APPLICATION TO MULTIPLICATIVE ANALOGUE OF C-CLASS FUNCTIONS, JP Journal of Fixed Point Theory and Applications, Volume 12, Number 1, 2017, Pages 1-35

Chapter 2

Fixed Point Results for Implicit Cyclic Weak φ-Multiplicative Contractions on Multiplicative Partial Metric Spaces

2.1 Brief Summary

> **Abstract B.1 1**
>
> In this chapter, we define the multiplicative version of cyclic weak φ-contractions implicitly via the C-class function of Ampadu and Ansari [Clement Ampadu and Arslan Hojat Ansari, FIXED POINT THEOREMS IN COMPLETE MULTIPLICATIVE METRIC SPACES WITH APPLICATION TO MULTIPLICATIVE ANALOGUE OF C-CLASS FUNCTIONS, JP Journal of Fixed Point Theory and Applications, August 2016, Volume 11, Issue 2, Pages 113-124]. Consequently, we obtain some fixed point results for such mappings including a Maia type theorem. Our results can be considered multiplicative generalization of certain results contained in [W.A. Kirk, P.S. Srinivasan and P. Veeramani, Fixed points for mappings satisfying cyclical contractive conditions, Fixed Point Theory. 4 (2003) 79-89]

2.2 Preliminaries

Taking inspiration from [S.G. Matthews, Partial metric topology, in: Proc. 8th Summer Conference on General Topology and Applications, Ann. New York Acad. Sci. 728 (1994) 183-197; S. Oltra and O. Valero, Banach's fixed point theorem for partial metric spaces, Rend. Istit.Mat. Univ. Trieste. 36 (2004) 17-26; S. Romaguera, A Kirk type characterization of completeness for partial metric spaces, Fixed Point Theory Appl. (2010), Article ID 493298; O. Valero, On Banach fixed point theorems for partial metric spaces, Appl. Gen. Topol. 6 (2005) 229-240] we introduce some concepts in multiplicative generalization of partial metric spaces that would be useful in the sequel.

Definition B.1 1

A multiplicative partial metric on a nonempty set X will be a function $p_m : X \times X \mapsto [1, \infty)$ such that for all $x, y, z \in X$ the following hold

(a) $x = y$ iff $p_m(x, x) = p_m(x, y) = p_m(y, y)$

(b) $p_m(x, x) \leq p_m(x, y)$

(c) $p_m(x, y) = p_m(y, x)$

(d) $p_m(x, y) \leq \frac{p_m(x,z) \cdot p_m(z,y)}{p_m(z,z)}$

Definition B.2 1

A multiplicative partial metric space will be a pair (X, p_m) such that X is a nonempty set and p_m is a multiplicative partial metric on X

Remark B.3 1

If $p_m(x, y) = 1$, then from (a) and (b) of Definition 1.1, it follows that $x = y$, but if $x = y$, $p_m(x, y)$ may not be one

Example B.4 1

A basic example of a multiplicative partial metric space is the pair $([1, \infty), p_m)$, where $p_m(x, y) = \max\{x, y\}$ for all $x, y \in [1, \infty)$

Remark B.5 1

Each multiplicative partial metric p_m on X generates a T_0 topology τ_{p_m} on X which has as a base the family of open p_m-balls $\{B_{p_m}(x, \epsilon) : x \in X, \epsilon > 1\}$, where $B_{p_m}(x, \epsilon) = \{y \in X : p_m(x, x) < \epsilon \cdot p_m(x, x)\}$ for all $x \in X$ and $\epsilon > 1$

Definition B.6 1

Let (X, p_m) be a multiplicative partial metric space

(a) We say $\{x_n\}$ in (X, p_m) multiplicative converges to a point $x \in X$ iff $p_m(x, x) = \lim_{n \to \infty} p_m(x, x_n)$

(b) We say $\{x_n\}$ in (X, p_m) is a multiplicative Cauchy sequence if $\lim_{n,k \to \infty} p_m(x_n, x_k) < \infty$

(c) We say (X, p_m) is multiplicative complete if every multiplicative Cauchy sequence $\{x_n\}$ in X multiplicative converges, with respect to τ_{p_m}, to a point $x \in X$ such that $p_m(x, x) = \lim_{n,k \to \infty} p_m(x_n, x_k)$

(d) A sequence $\{x_n\}$ in (X, p_m) will be called multiplicative 0-Cauchy if $\lim_{n,k \to \infty} p_m(x_n, x_k) = 1$. We say that (X, p_m) is multiplicative 0-complete if every multiplicative 0-Cauchy sequence in X multiplicative converges, with respect to τ_{p_m}, to a point $x \in X$ such that $p_m(x, x) = 1$

Remark B.7 1

Define $p_m(x, A) = \inf\{p_m(x, a) : a \in A\}$, then $a \in \overline{A}$ iff $p_m(a, A) = p_m(a, a)$, where $\overline{A}$ denotes the closure of A

From now on we introduce some concepts related to cyclic representation that would be useful in the sequel

> **Definition B.8 1**
>
> Let X be a nonempty set, k a positive integer, and $T: X \mapsto X$ be a mapping. A finite family $A_1, \cdots, A_k$ of nonempty subsets of X is a cyclic representation of X with respect to T if
>
> (i) $\bigcup_{i=1}^{k} A_i = X$
>
> (ii) $T(A_1) \subset A_2, T(A_2) \subset A_3, \cdots, T(A_k) \subset A_1$

> **Definition B.9 1**
>
> Let (X, p_m) be a multiplicative partial metric space, k a positive integer, $A_1, \cdots, A_k$ be closed nonempty subsets of X, and $Y = \bigcup_{i=1}^{k} A_i$. We say $T: Y \mapsto Y$ is an implicit cyclic weak φ- multiplicative contraction if
>
> (i) $A_1, \cdots, A_k$ is a cyclic representation of Y with respect to T
>
> (ii) there exists a nondecreasing function $\varphi : [1, \infty) \mapsto [1, \infty)$ with $\varphi(t) > 1$ for $t > 1$ and $\varphi(1) = 1$, such that
>
> $$p_m(Tx, Ty) \leq F\left[\frac{p_m(x,y)}{\varphi(p_m(x,y))}\right]$$
>
> for all $x \in A_i$ and $y \in A_{i+1}$, $i = 1, 2, \cdots, k$, where $A_{k+1} = A_1$, and $F(x,y) := F\left(\frac{x}{y}\right)$ is a multiplicative C-class function [Clement Ampadu and Arslan Hojat Ansari, FIXED POINT THEOREMS IN COMPLETE MULTIPLICATIVE METRIC SPACES WITH APPLICATION TO MULTIPLICATIVE ANALOGUE OF C-CLASS FUNCTIONS, JP Journal of Fixed Point Theory and Applications, August 2016, Volume 11, Issue 2, Pages 113-124]

From now on we introduce some notations that would be useful in the sequel

> **Notation B.10 1**
>
> Φ will denote the class of all nondecreasing functions $\varphi : [1, \infty) \mapsto [1, \infty)$ continuous at one, such that $\varphi(1) = 1$ and $\varphi(t) > 1$ for all $t > 1$

> **Notation B.11 1**
>
> Let $T: X \mapsto X$, where X is a nonempty set. $Fix(T)$ will denote the set $\{x \in X : x = Tx\}$

> **Notation B.12 1**
>
> Ψ will denote the class of all functions $\psi : [1, \infty) \mapsto [1, \infty)$ such that $\frac{t}{\psi(t)} := \varphi(t) \in \Phi$

2.3 Main Results

> **Lemma B.1 1**
>
> Let (X, p_m) be a multiplicative partial metric space and $\{x_n\} \subset X$. If $x_n \to x \in X$ and $p_m(x,x) = 1$, then, $\lim_{n \to \infty} p_m(x_n, z) = p_m(x, z)$ for all $z \in X$

CHAPTER 2. FIXED POINT RESULTS FOR IMPLICIT CYCLIC WEAK φ-MULTIPLICATIVE CONTRACTIONS ON MULTIPLICATIVE PARTIAL METRIC SPACES

Proof of Lemma B.1 1

By the multiplicative triangle inequality,

$$\frac{p_m(x,z)}{p_m(x_n,x)} \leq p_m(x_n,z) \leq p_m(x,z) \cdot p_m(x_n,x)$$

If we take limits in the above as $n \to \infty$, we deduce that $\lim_{n\to\infty} p_m(x_n,z) = p_m(x,z)$.

Lemma B.2 1

Let (X, p_m) be a multiplicative partial metric space, k a positive integer, $A_1, \cdots, A_k$ be closed nonempty subsets of X and $Y = \bigcup_{i=1}^{k} A_i$. If $T : Y \mapsto Y$ is an implicit cyclic weak φ-multiplicative contraction, then

(a) $p_m(Tx, Ty) \leq p_m(x,y)$ for all $x \in A_i$ and $y \in A_{i+1}$, $i = 1, \cdots, k$

(b) $p_m(T^n x, T^{n+1} y) \to 1$ for all $x, y \in A_i$, $i = 1, \cdots, k$

(c) $p_m(T^{k(n+1)} x, T^{kn} x) \to 1$ for all $x \in A_i$, $i = 1, \cdots, k$

(d) If $z \in Fix(T)$, then $p_m(z,z) = 1$

Proof of Lemma B.2 1

Only properties (b) and (d) are nontrivial. First, we prove (b). Let $x, y \in A_i$ and define $t_n = p_m(T^n x, T^{n+1} y)$. Since T is an implicit cyclic weak φ-multiplicative contraction, we have by the properties of F, for all $n \in \mathbb{N}$,

$$t_{n+1} \leq F\left[\frac{t_n}{\varphi(t_n)}\right] \leq t_n$$

Thus the sequence $\{t_n\}$ is nonincreasing and hence there exists $\alpha \geq 1$ such that $t_n \to \alpha$. We claim that $\alpha = 1$. Assume $\alpha > 1$, then there exist $n_0 \in \mathbb{N}$ such that $t_1 < \varphi(\alpha)^n$ for all $n \geq n_0$. Now by the monotonicity of φ for all $n \geq n_0$, we have,

$$t_{n+1} \leq F\left[\frac{t_n}{\varphi(\alpha)}\right] \leq F\left[\frac{t_{n-1}}{\varphi(\alpha)^2}\right] \leq \cdots \leq F\left[\frac{t_1}{\varphi(\alpha)^n}\right]$$

which is a contradiction and so $\alpha = 1$. Property (d) follows from

$$p_m(z,z) = p_m(Tz, Tz) \leq F\left[\frac{p_m(z,z)}{\varphi(p_m(z,z))}\right]$$

which is possible only if $p_m(z,z) = 1$

Lemma B.3 1

Let (X, p_m) be a multiplicative partial metric space, k a positive integer, $A_1, \cdots, A_k$ be closed nonempty subsets of X and $Y = \bigcup_{i=1}^{k} A_i$. If $T : Y \mapsto Y$ is an implicit cyclic weak φ-multiplicative contraction, given $x_0 \in A_i$ ($i = 1, 2, \cdots, k$), then for every $\epsilon > 1$ there exists n_ϵ such that $p_m(T^{ks} x_0, T^{kn+1} x_0) < \epsilon$ for all $s > n \geq n_\epsilon$

Proof of Lemma B.3 1

Suppose not, then there exist $\epsilon > 1$ such that for each $q \geq 1$, there exists $s_q > n_q \geq k$ so that
$$p_m(T^{ks_q}x_0, T^{kn_q+1}x_0) \geq \epsilon$$
and
$$p_m(T^{k(s_q-1)}x_0, T^{kn_q+1}x_0) < \epsilon$$

Now observe that
$$\epsilon \leq p_m(T^{ks_q}x_0, T^{kn_q+1}x_0)$$
$$\leq p_m(T^{ks_q}x_0, T^{k(s_q-1)}x_0) \cdot p_m(T^{k(s_q-1)}x_0, T^{kn_q+1}x_0)$$
$$\leq p_m(T^{ks_q}x_0, T^{k(s_q-1)}x_0) \cdot \epsilon$$

By the previous Lemma, if we take limits in the inequality immediately above, we deduce that $\lim_{k \to \infty} p_m(T^{ks_q}x_0, T^{kn_q+1}x_0) = \epsilon$. Now observe by the previous Lemma, that
$$p_m(T^{k(s_q+1)}x_0, T^{k(n_q+1)+1}x_0) \leq p_m(T^{ks_q+1}x_0, T^{kn_q+2}x_0)$$

it follows that we have the following

$$p_m(T^{ks_q}x_0, T^{kn_q+1}x_0) \leq p_m(T^{ks_q}x_0, T^{k(s_q+1)}x_0) \cdot p_m(T^{k(s_q+1)}x_0, T^{k(n_q+1)+1}x_0)$$
$$\cdot p_m(T^{k(n_q+1)+1}x_0, T^{kn_q+1}x_0)$$
$$\leq p_m(T^{ks_q}x_0, T^{k(s_q+1)}x_0) \cdot p_m(T^{ks_q+1}x_0, T^{kn_q+2}x_0)$$
$$\cdot p_m(T^{k(n_q+1)+1}x_0, T^{kn_q+1}x_0)$$
$$\leq p_m(T^{ks_q}x_0, T^{k(s_q+1)}x_0) \cdot F\left[\frac{p_m(T^{ks_q}x_0, T^{kn_q+1}x_0)}{\varphi(\epsilon)}\right]$$
$$\cdot p_m(T^{k(n_q+1)+1}x_0, T^{kn_q+1}x_0)$$

By the previous Lemma, if we take limits in the equality immediately above as $k \to \infty$, we deduce that
$$\epsilon \leq F\left[\frac{\epsilon}{\varphi(\epsilon)}\right]$$

which is a contradiction. Consequently, for every $\epsilon > 1$, there exists n_ϵ such that $p_m(T^{ks}x_0, T^{kn+1}x_0) < \epsilon$ for all $s > n \geq n_\epsilon$

Lemma B.4 1

Let (X, p_m) be a multiplicative partial metric space, k a positive integer, $A_1, \cdots, A_k$ be closed nonempty subsets of X, and $Y = \bigcup_{i=1}^{k} A_i$, and $T : Y \mapsto Y$ be an implicit cyclic weak φ- multiplicative contraction. Assuming that there exist a sequence $\{y_n\} \subset Y$ such that $p_m(y_n, Ty_n) \to 1$ as $n \to \infty$ and $z \in Fix(T)$, then $\lim_{n \to \infty} y_n = z$. Moreover, T has at most one fixed point.

Proof of Lemma B.4 1

Assume that the sequence $\{y_n\}$ doesn't multiplicative converge to z, then $\limsup_{n \to \infty} p_m(y_n, z) = \alpha > 1$. Let $N = \{n : p_m(y_n, Ty_n) < \varphi(\sqrt{\alpha}) \text{ and } p_m(y_n, z) > \sqrt{\alpha}\}$. For all $n \in N$, we have,

$$p_m(y_n, z) \leq \frac{p_m(y_n, Ty_n) \cdot p_m(Ty_n, Tz)}{p_m(Ty_n, Ty_n)}$$
$$\leq p_m(y_n, Ty_n) \cdot F\left[\frac{p_m(y_n, z)}{\varphi(\sqrt{\alpha})}\right]$$
$$< p_m(y_n, z)$$

which is a contradiction, and so the sequence $\{y_n\}$ multiplicative converges to z. Lemma B.2 ensures that there exist $\{y_n\} \subset Y$ such that $\lim_{n \to \infty} p_m(y_n, Ty_n) = 1$. We now show that T has at most one fixed point. Suppose not, let $w \in Fix(T)$. Now observe that,

$$p_m(z, w) \leq \frac{p_m(y, z_n) \cdot p_m(y_n, w)}{p_m(y_n, y_n)}$$
$$\leq p_m(y, z_n) \cdot p_m(y_n, w)$$

Since $p_m(y, z_n), p_m(y_n, w) \to 1$ as $n \to \infty$. If we take limits in the above inequality, we deduce that $p_m(z, w) \leq 1$, and hence, $z = w$

Theorem B.5 1

Let (X, p_m) be a multiplicative partial metric space, k a positive integer, $A_1, \cdots, A_k$ be multiplicative 0-complete nonempty subsets of X and $Y = \bigcup_{i=1}^{k} A_i$. If $T : Y \mapsto Y$ is an implicit cyclic weak φ-multiplicative contraction, then T has a unique fixed point $z \in \bigcap_{i=1}^{k} A_i$

Proof of Theorem B.5 1

Let $x_0 \in Y := \bigcup_{i=1}^{k} A_i$ and $\epsilon > 1$. Let $\{x_n\}$ be the Picard iteration defined by $x_n = Tx_{n-1}$ for all n. By Lemma B.2 and Lemma B.3, there exists n_ϵ such that $p_m(x_{kn+1}, x_{kn}) < \sqrt{\epsilon}$ and $p_m(x_{ks}, x_{kn+1}) < \sqrt{\epsilon}$ for all $s > n \geq n_\epsilon$. Now observe that

$$p_m(x_{ks}, x_{kn}) \leq \frac{p_m(x_{ks}, x_{kn+1}) \cdot p_m(x_{kn+1}, x_{kn})}{p_m(x_{kn+1}, x_{kn+1})}$$
$$\leq p_m(x_{ks}, x_{kn+1}) \cdot p_m(x_{kn+1}, x_{kn})$$
$$< \epsilon$$

for all $s > n \geq n_\epsilon$. Consequently, $\lim_{s,n \to \infty} p_m(x_{ks}, x_{kn}) = 1$, and hence $\{x_{kn}\}$ is a multiplicative 0-Cauchy sequence. By Lemma B.2, $\{x_n\}$ is also a multiplicative 0-Cauchy sequence. Since Y is multiplicative 0-complete, there exist $z \in Y$ such that $p_m(x_n, z) \to p_m(z, z) = 1$. Also $x_{kn+i} \to z$, for $i = 0, 1, 2, \cdots, k-1$. This implies that $z \in \bigcap_{i=1}^{k} A_i$, since each A_i is multiplicative 0-complete. We show that $z = Tz$. Observe that

$$p_m(z, Tz) \leq \frac{p_m(z, x_{n+1}) \cdot p_m(Tx_n, Tz)}{p_m(x_{n+1}, x_{n+1})}$$
$$\leq p_m(z, x_{n+1}) \cdot F\left[\frac{p_m(x_n, z)}{\varphi(p_m(x_n, z))}\right]$$

Since $\varphi(p_m(x_n, z)) \to 1$ as $n \to \infty$, if we take limits in the above inequality we deduce that $p_m(z, Tz) \leq 1$, and thus $z = Tz$. The uniqueness of the fixed point is obvious

Theorem B.6 1

Let (X, p_m) be a multiplicative partial metric space, k a positive integer, $A_1, \cdots, A_k$ be multiplicative 0-complete nonempty subsets of X, $Y = \bigcup_{i=1}^{k} A_i$, and $T : Y \mapsto Y$ be an implicit cyclic weak φ-multiplicative contraction. Assuming there exists a sequence $\{y_n\} \subset Y$ such that $p_m(y_n, y) \to p_m(y, y) = 1$ and $p_m(y_{n+1}, Ty_n) \to 1$ as $n \to \infty$, then for all $x \in Y$ we have $\lim_{n \to \infty} p_m(y_n, T^n x) = 1$

Proof of Theorem B.6 1

By the previous theorem, T has a unique fixed point z such that $p_m(z, z) = 1$. Now by Lemma B.1, $\lim_{n \to \infty} p_m(y_n, z) = p_m(y, z)$. If $y \neq z$, then $p_m(y, z) > 1$, and thus there exists N such that $p_m(y_n, z) \geq \sqrt{p_m(y, z)}$ for all $n \geq N$. Now observe that for all $n \geq N$ we have

$$p_m(y_{n+1}, z) \leq \frac{p_m(y_{n+1}, Ty_n) \cdot p_m(Ty_n, Tz)}{p_m(Ty_n, Ty_n)}$$

$$\leq p_m(y_{n+1}, Ty_n) \cdot F\left[\frac{p_m(y_n, z)}{\varphi(\sqrt{p_m(y, z)})}\right]$$

Now taking limits in the above as $n \to \infty$, we deduce that $p_m(y, z) \leq F\left[\frac{p_m(y,z)}{\varphi(\sqrt{p_m(y,z)})}\right]$ which is possible only if $p_m(y, z) = 1$, that is, if $y = z$. On the other hand, for all $x \in Y$, by Lemma B.2 and Lemma B.4, we have

$$\lim_{n \to \infty} p_m(y_{n+1}, T^n x) \leq \lim_{n \to \infty} [p_m(y_{n+1}, z) \cdot p_m(z, T^n z)] = 1$$

If we take $F(x, y) := F(\frac{x}{y}) = \frac{x}{y}$, and define $\varphi : [1, \infty) \mapsto [1, \infty)$ by $\varphi(t) = t^{1-v}$, where $v \in (0, 1)$, then we get the multiplicative version of Theorem 1.3 [W.A. Kirk, P.S. Srinivasan and P. Veeramani, Fixed points for mappings satisfying cyclical contractive conditions, Fixed Point Theory. 4 (2003) 79-89] from Theorem B.5 above as follows

Corollary B.7 1

Let $A_1, \cdots, A_k$ be a finite family of nonempty closed subsets of a complete multiplicative metric space (X, m), and suppose $T : \bigcup_{i=1}^{k} A_i \mapsto \bigcup_{i=1}^{k} A_i$ satisfies the following conditions

(a) $T(A_1) \subset A_2, T(A_2) \subset A_3, \cdots, T(A_k) \subset A_1$

(b) there exists $v \in (0, 1)$ such that $m(Tx, Ty) \leq m(x, y)^v$ for all $x \in A_i$ and $y \in A_{i+1}$ for $1 \leq i \leq k$, where $A_{k+1} = A_1$

Then T has a unique fixed point

If we take $F(x, y) := F(\frac{x}{y}) = \frac{x}{y}$, and let $\psi \in \Psi$ be a function, then there exist $\varphi \in \Phi$ such that $\frac{t}{\psi(t)} = \varphi(t)$, thus we get the multiplicative version of a result contained in [D.W. Boyd and J.S.W.Wong, On nonlinear contractions, Proc. Amer. Math. Soc. 20 (1969)458-464] as follows

Corollary B.8 1

Let $A_1, \cdots, A_k$ be a finite family of nonempty closed subsets of a complete multiplicative metric space (X, m), and suppose $T : \bigcup_{i=1}^{k} A_i \mapsto \bigcup_{i=1}^{k} A_i$ satisfies the following conditions

(a) $T(A_1) \subset A_2, T(A_2) \subset A_3, \cdots, T(A_k) \subset A_1$

(b) there exists $\psi \in \Psi$ such that $m(Tx, Ty) \leq \psi(m(x, y))$ for all $x \in A_i$ and $y \in A_{i+1}$ for $1 \leq i \leq k$, where $A_{k+1} = A_1$

Then T has a unique fixed point

Finally the Maia type result for implicit cyclic φ-multiplicative contractions is obtained as follows

> **Theorem B.9 1**
>
> Let X be a nonempty set, p_1 and p_2 be two multiplicative partial metrics on X, k a positive integer, $A_1, \cdots, A_k$ be closed nonempty subsets of (X, p_1), $Y = \bigcup_{i=1}^{k} A_i$ and $T : Y \mapsto Y$. Assuming that
>
> (i) $A_1, \cdots, A_k$ is a cyclic representation of Y with respect to T
>
> (ii) $p_1(x, y) \leq p_2(x, y)$, for any $x, y \in Y$
>
> (iii) (Y, p_1) is a multiplicative 0-complete multiplicative partial metric space
>
> (iv) $T : (Y, p_1) \mapsto (Y, p_1)$ is continuous
>
> (v) $T : (Y, p_2) \mapsto (Y, p_2)$ is an implicit cyclic weak φ-multiplicative contraction
>
> Then T has a unique fixed point

2.4 Exercises

> **Exercise B.1 1**
>
> Taking inspiration from [Cristina Di Bari and Pasquale Vetro, Fixed points for weak φ-contractions on partial metric spaces, Int. Jr. of Engineering, Contemporary Mathematics and Sciences, Vol. 1, No. 1, January-June 2015, Copyright: MUK Publications, ISSN No: 2250-3099] give an example of an implicit cyclic weak φ- multiplicative contraction

> **Exercise B.2 1**
>
> Let $\{A_i\}_{i=1}^{k}$ be nonempty closed subsets of a multiplicative partial metric space (X, p_m), and suppose $T : \bigcup_{i=1}^{k} A_i \mapsto \bigcup_{i=1}^{k} A_i$ is a cyclic operator, that is, T satisfies $T(A_i) \subseteq A_{i+1}$ for all $i \in \{1, 2, \cdots, k\}$, where $A_{k+1} = A_1$. We will say T is an implicit $(\psi - \phi)$-Kannan type multiplicative contraction if there exists nonnegative constants α, β with $0 < \alpha + \beta \leq 1$, $\alpha > 0$ such that for any $x \in A_i$, $y \in A_{i+1}$, $i = 1, 2, \cdots, k$, we have
>
> $$\psi(p_m(Tx, Ty)) \leq F\left[\frac{\psi(p_m(x, Tx)^\alpha \cdot p_m(y, Ty)^\beta)}{\phi(p_m(x, Tx), p_m(y, Ty))}\right]$$
>
> where $\psi : [1, \infty) \mapsto [1, \infty)$ is continuous and nondecreasing, and $\psi(t) = 1$ iff $t = 1$, $\phi : [1, \infty)^2 \mapsto [1, \infty)$ is a continuous function such that $\phi(a, b) = 1$ iff $a = b = 1$, and $F(x, y) := F(\frac{x}{y})$ is a multiplicative C-class function [Clement Ampadu and Arslan Hojat Ansari, FIXED POINT THEOREMS IN COMPLETE MULTIPLICATIVE METRIC SPACES WITH APPLICATION TO MULTIPLICATIVE ANALOGUE OF C-CLASS FUNCTIONS, JP Journal of Fixed Point Theory and Applications, August 2016, Volume 11, Issue 2, Pages 113-124].
>
> Taking inspiration from [W.B. DOMI, S. AL-SHARIF, H. ALMEFLEH, NEW RESULTS ON CYCLIC NONLINEAR CONTRACTIONS IN PARTIAL METRIC SPACES, TWMS J. App. Eng. Math. V.5, N.2, 2015, pp.158-168] prove the following:
>
> Let $\{A_i\}_{i=1}^{k}$ be nonempty closed subsets of a multiplicative complete multiplicative partial metric space and $T : \bigcup_{i=1}^{k} A_i \mapsto \bigcup_{i=1}^{k} A_i$ be an implicit $(\psi - \phi)$-Kannan type multiplicative contraction, then T has a unique fixed point $z \in \bigcap_{i=1}^{k} A_i$

Exercise B.3 1

Let $\{A_i\}_{i=1}^k$ be nonempty closed subsets of a multiplicative partial metric space (X, p_m), and suppose $T : \bigcup_{i=1}^k A_i \mapsto \bigcup_{i=1}^k A_i$ is a cyclic operator, that is, T satisfies $T(A_i) \subseteq A_{i+1}$ for all $i \in \{1, 2, \cdots, k\}$, where $A_{k+1} = A_1$. We will say T is an implicit $(\psi - \phi)$-Chatterjea type multiplicative contraction if there exists nonnegative constants α, β with $0 < \alpha + \beta < 1$, $0 < \alpha \leq \beta$ such that for any $x \in A_i, y \in A_{i+1}, i = 1, 2, \cdots, k$, we have

$$\psi(p_m(Tx, Ty)) \leq F\left[\frac{\psi(p_m(x, Ty)^\alpha \cdot p_m(y, Tx)^\beta)}{\phi(p_m(x, Ty), p_m(y, Tx))}\right]$$

where $\psi : [1, \infty) \mapsto [1, \infty)$ is continuous and nondecreasing, and $\psi(t) = 1$ iff $t = 1$, $\phi : [1, \infty)^2 \mapsto [1, \infty)$ is a continuous function such that $\phi(a, b) = 1$ iff $a = b = 1$, and $F(x, y) := F(\frac{x}{y})$ is a multiplicative C-class function [Clement Ampadu and Arslan Hojat Ansari, FIXED POINT THEOREMS IN COMPLETE MULTIPLICATIVE METRIC SPACES WITH APPLICATION TO MULTIPLICATIVE ANALOGUE OF C-CLASS FUNCTIONS, JP Journal of Fixed Point Theory and Applications, August 2016, Volume 11, Issue 2, Pages 113-124].

Taking inspiration from [W.B. DOMI, S. AL-SHARIF, H. ALMEFLEH, NEW RESULTS ON CYCLIC NONLINEAR CONTRACTIONS IN PARTIAL METRIC SPACES, TWMS J. App. Eng. Math. V.5, N.2, 2015, pp.158-168] prove the following:

Let $\{A_i\}_{i=1}^k$ be nonempty closed subsets of a multiplicative complete multiplicative partial metric space and $T : \bigcup_{i=1}^k A_i \mapsto \bigcup_{i=1}^k A_i$ be an implicit $(\psi - \phi)$-Chatterjea type multiplicative contraction, then T has a unique fixed point $z \in \bigcap_{i=1}^k A_i$

Exercise B.4 1

Multiplicative integral was introduced in [A. E. Bashirov, E. M. Kurpinar and A. Ozyapici, Multiplicative calculus and its applications, J. Math. Anal. Appl. 337(1) (2008), 36-48] and we gave an equivalent interpretation [Clement Ampadu, Arslan Hojat Ansari and Memudu Olaposi Olatinwo, FIXED POINT THEOREMS USING MULTIPLICATIVE CONTRACTIVE DEFINITIONS WITH APPLICATION TO MULTIPLICATIVE ANALOGUE OF C-CLASS FUNCTIONS, JP Journal of Fixed Point Theory and Applications, Volume 12, Number 1, 2017, Pages 1-35]. Let ρ be a Lebesgue integrable mapping and define $\psi : [1, \infty) \mapsto [1, \infty)$ by $\psi(t) = \int_1^t \rho(s) ds > 1$ for $t > 1$. In addition, let $\phi = 1$ and $F(x, y) := F(\frac{x}{y}) = \frac{x}{y}$. Under these conditions

(a) State the Corollary arising from Exercise B.2

(b) State the Corollary arising from Exercise B.3

2.5 References

(1) W.A. Kirk, P.S. Srinivasan and P. Veeramani, Fixed points for mappings satisfying cyclical contractive conditions, Fixed Point Theory. 4 (2003) 79-89

(2) S.G. Matthews, Partial metric topology, in: Proc. 8th Summer Conference on General Topology and Applications, Ann. New York Acad. Sci. 728 (1994)183-197

(3) S. Oltra and O. Valero, Banach's fixed point theorem for partial metric spaces, Rend. Istit.Mat. Univ. Trieste. 36 (2004) 17-26

(4) S. Romaguera, A Kirk type characterization of completeness for partial metric spaces, Fixed Point Theory Appl. (2010), Article ID 493298

(5) O. Valero, On Banach fixed point theorems for partial metric spaces, Appl. Gen. Topol. 6 (2005) 229-240

(6) D.W. Boyd and J.S.W.Wong, On nonlinear contractions, Proc. Amer. Math. Soc. 20 (1969)458-464

(7) Cristina Di Bari and Pasquale Vetro, Fixed points for weak φ-contractions on partial metric spaces, Int. Jr. of Engineering, Contemporary Mathematics and Sciences, Vol. 1, No. 1, January-June 2015, Copyright: MUK Publications, ISSN No: 2250-3099

(8) W.B. DOMI, S. AL-SHARIF, H. ALMEFLEH, NEW RESULTS ON CYCLIC NONLINEAR CONTRACTIONS IN PARTIAL METRIC SPACES, TWMS J. App. Eng. Math. V.5, N.2, 2015, pp.158-168

(9) A. E. Bashirov, E. M. Kurpinar and A. Ozyapici, Multiplicative calculus and its applications, J. Math. Anal. Appl. 337(1) (2008), 36-48

(10) Clement Ampadu, Arslan Hojat Ansari and Memudu Olaposi Olatinwo, FIXED POINT THEOREMS USING MULTIPLICATIVE CONTRACTIVE DEFINITIONS WITH APPLICATION TO MULTIPLICATIVE ANALOGUE OF C-CLASS FUNCTIONS, JP Journal of Fixed Point Theory and Applications, Volume 12, Number 1, 2017, Pages 1-35

Chapter 3

Existence and Uniqueness of Fixed Points for Implicit Almost Generalized Cyclic (ψ, φ)-Weak Multiplicative Contractive Type Mappings

3.1 Brief Summary

Abstract C.1 1

In this chapter we define the multiplicative version of almost generalized cyclic (ϕ, φ)-weak contractive mappings implicitly via the C-class function of Ampadu and Ansari [Clement Ampadu and Arslan Hojat Ansari, FIXED POINT THEOREMS IN COMPLETE MULTIPLICATIVE METRIC SPACES WITH APPLICATION TO MULTIPLICATIVE ANALOGUE OF C-CLASS FUNCTIONS, JP Journal of Fixed Point Theory and Applications, August 2016, Volume 11, Issue 2, Pages 113-124]. Consequently, we obtain some existence and uniqueness results for such mappings.

3.2 Preliminaries

Definition C.1 1

[W. A. Kirk, P. S. Srinivasan, and P. Veeramani, "Fixed points for mappings satisfying cyclical contractive conditions," Fixed Point Theory, vol. 4, no. 1, pp. 79–89, 2003] Let X be a nonempty set, k a positive integer, and $T : X \mapsto X$ be a mapping. $X = \bigcup_{i=1}^{k} A_i$ is said to be a cyclic representation of X with respect to T if

(a) A_i, $i = 1, 2, \cdots, k$ are nonempty closed sets

(b) $T(A_1) \subseteq A_2, \cdots, T(A_{k-1}) \subseteq A_k, T(A_k) \subseteq A_1$

Notation C.2 1

Φ will denote the class of all functions $\varphi : [1, \infty) \mapsto [1, \infty)$ satisfying

(a) φ is lower semi-continuous

(b) $\varphi^{-1}(\{1\}) = \{1\}$

> **Notation C.3 1**
>
> Ψ will denote the class of all continuous functions $\psi : [1, \infty) \mapsto [1, \infty)$

> **Definition C.4 1**
>
> Let (X, m) be a multiplicative metric space. Let k be a positive integer and let $A_1, \cdots, A_k$ be nonempty subsets of X and $Y = \bigcup_{i=1}^{k} A_i$. The mapping $T : Y \mapsto Y$ will be called an implicit almost generalized cyclic (ψ, φ)-weak multiplicative contraction, if the following holds
>
> (a) $Y = \bigcup_{i=1}^{k} A_i$ is a cyclic representation of Y with respect to T
>
> (b) there exists $L \geq 0$, $\psi \in \Psi$ and $\varphi \in \Phi$ such that
>
> $$\psi(m(Tx, Ty)) \leq F\left[\frac{\psi(Q(x,y)) \cdot \min\{m(x, Tx), m(y, Ty), m(x, Ty), m(y, Tx)\}^L}{\varphi(Q(x,y))}\right]$$
>
> for all $x, y \in A_i \times A_{i+1}$, $i = 1, 2, \cdots, k$ (with $A_{k+1} = A_1$), where
>
> $$Q(x, y) = \max\{m(x, y), m(Tx, x), m(Ty, y), \sqrt{m(x, Ty) \cdot m(y, Tx)}\}$$
>
> and $F(x, y) := F(\frac{x}{y})$ is a multiplicative C-class function [Clement Ampadu and Arslan Hojat Ansari, FIXED POINT THEOREMS IN COMPLETE MULTIPLICATIVE METRIC SPACES WITH APPLICATION TO MULTIPLICATIVE ANALOGUE OF C-CLASS FUNCTIONS, JP Journal of Fixed Point Theory and Applications, August 2016, Volume 11, Issue 2, Pages 113-124]

3.3 Main Results

> **Theorem C.1 1**
>
> Let $\{A_i\}_{i=1}^{k}$ be nonempty closed subsets of a complete multiplicative metric space (X, m) and $Y = \bigcup_{i=1}^{k} A_i$. Let $T : Y \mapsto Y$ be an implicit almost generalized cyclic (ψ, φ)-weak multiplicative contraction mapping. Then T has a unique fixed point that belongs to $\bigcap_{i=1}^{k} A_i$

> **Proof of Theorem C.1 1**

Let $x_0 \in A_1$ (such a point exists since $A_1 \neq \emptyset$). Define the sequence $\{x_n\}$ in X by $x_{n+1} = Tx_n$ for $n = 0, 1, 2, \cdots$. We show that $\lim_{n \to \infty} m(x_n, x_{n+1}) = 1$. If for some j we have $x_{j+1} = x_j$, then $\lim_{n \to \infty} m(x_n, x_{n+1}) = 1$ is immediate. So we assume that $m(x_n, x_{n+1}) > 1$ for all n. From Definition C.4(a), we observe that for all n, there exists $i = i(n) \in \{1, 2, \cdots, k\}$ such that $(x_n, x_{n+1}) \in A_i \times A_{i+1}$. From Definition C.4(b), we have for $n = 1, 2, \cdots$

$$\psi(m(x_n, x_{n+1})) \leq F\left[\frac{\psi(Q(x_{n-1}, x_n))}{\varphi(Q(x_{n-1}, x_n))}\right]$$

where

$$Q(x_{n-1}, x_n) = \max\{m(x_{n-1}, x_n), m(Tx_{n-1}, x_{n-1}), m(Tx_n, x_n),$$
$$\sqrt{m(x_{n-1}, Tx_n) \cdot m(x_n, Tx_{n-1})}\}$$
$$= \max\{m(x_{n-1}, x_n), m(x_n, x_{n-1}), m(x_{n+1}, x_n), \sqrt{m(x_{n-1}, x_{n+1})}\}$$
$$= \max\{m(x_{n-1}, x_n), m(x_{n+1}, x_n)\}$$

Now suppose $Q(x_{n-1}, x_n) = m(x_{n+1}, x_n)$, then from $\psi(m(x_n, x_{n+1})) \leq F\left[\frac{\psi(Q(x_{n-1}, x_n))}{\varphi(Q(x_{n-1}, x_n))}\right]$, we deduce that $\psi(m(x_n, x_{n+1})) \leq F\left[\frac{\psi(m(x_{n+1}, x_n))}{\varphi(m(x_{n+1}, x_n))}\right]$, from which it follows that $\varphi(m(x_{n+1}, x_n)) = 1$ and hence $m(x_{n+1}, x_n) = 1$, a contradiction with our assumption that $m(x_n, x_{n+1}) > 1$ for all n. Thus, we have $Q(x_{n-1}, x_n) = m(x_{n-1}, x_n)$, which implies that $\{m(x_n, x_{n+1})\}$ is a decreasing sequence of positive real numbers, hence there exists $r \geq 1$ such that $\lim_{n \to \infty} m(x_n, x_{n+1}) = r$. We claim that $r = 1$, if not, let us suppose that $r > 1$. Now using $\lim_{n \to \infty} m(x_n, x_{n+1}) = r$, the continuity of ψ and the lower semi-continuity of φ, if we take limits as $n \to \infty$ in

$$\psi(m(x_n, x_{n+1})) \leq F\left[\frac{\psi(Q(x_{n-1}, x_n))}{\varphi(Q(x_{n-1}, x_n))}\right]$$

we deduce that $\psi(r) \leq F\left[\frac{\psi(r)}{\varphi(r)}\right]$ which implies that $\varphi(r) = 1$, that is, $r = 1$, and thus $\lim_{n \to \infty} m(x_n, x_{n+1}) = 1$. Now we show that $\{x_n\}$ is a multiplicative Cauchy sequence in (X, m). Suppose $\{x_n\}$ is not a multiplicative Cauchy sequence. Then there exists $\epsilon > 1$ for which we can find two sequences of positive integers $\{v(j)\}$ and $\{n(j)\}$ such that for all positive integers j, $n(j) > v(j) > j$, $m(x_{v(j)}, x_{n(j)}) \geq \epsilon$, and $m(x_{v(j)}, x_{n(j)-1}) < \epsilon$. By the multiplicative triangle inequality, we deduce that

$$\epsilon \leq m(x_{v(j)}, x_{n(j)})$$
$$\leq m(x_{v(j)}, x_{n(j)-1}) \cdot m(x_{n(j)-1}, x_{n(j)})$$
$$< \epsilon \cdot m(x_{n(j)-1}, x_{n(j)})$$

Since $\lim_{n \to \infty} m(x_n, x_{n+1}) = 1$, if we take limits in the above inequality we deduce that $\lim_{j \to \infty} m(x_{v(j)}, x_{n(j)}) = \epsilon$. Now observe for all j, there exists $t(j) \in \{1, 2, \cdots, k\}$ such that $n(j) - v(j) + t(j) \equiv_* 1 \pmod{k}$. It follows that $x_{v(j)-t(j)}$ (for j large enough, v(j)>t(k)) and $x_{n(j)}$ lie in different adjacently labeled sets A_i and A_{i+1} for certain $i \in \{1, 2, \cdots, k\}$. Now observe from the multiplicative contraction condition of the theorem we have

$$\psi(m(x_{v(j)-t(j)+1}, x_{n(j)+1})) = \psi(m(Tx_{v(j)-t(j)}, Tx_{n(j)}))$$
$$\leq F\left[\frac{\psi(Q(x_{v(j)-t(j)}, x_{n(j)})) \cdot W(x_{v(j)-t(j)}, x_{n(j)})}{\varphi(Q(x_{v(j)-t(j)}, x_{n(j)}))}\right]$$

where

$$W(x_{v(j)-t(j)}, x_{n(j)}) = \min\{m(x_{v(j)-t(j)}, x_{v(j)-t(j)+1}), m(x_{n(j)}, x_{n(j)+1}), m(x_{v(j)-t(j)}, x_{n(j)+1}),$$
$$m(x_{n(j)}, x_{v(j)-t(j)+1})\}^L$$

On the other hand, observe that

$$Q(x_{v(j)-t(j)}, x_{n(j)}) = \max\{m(x_{v(j)-t(j)}, x_{n(j)}), m(x_{v(j)-t(j)+1}, x_{v(j)-t(j)}), m(x_{n(j)+1}, x_{n(j)}),$$
$$\sqrt{m(x_{v(j)-t(j)}, x_{n(j)+1}) \cdot m(x_{v(j)-t(j)+1}, x_{n(j)})}\}$$

Now observe by the multiplicative triangle inequality, as $j \to \infty$, we have

$$\frac{\max\{m(x_{v(j)-t(j)}, x_{n(j)})^2, m(x_{n(j)}, x_{v(j)})^2\}}{m(x_{v(j)-t(j)}, x_{n(j)}) \cdot m(x_{n(j)}, x_{v(j)})} \leq m(x_{v(j)-t(j)}, x_{v(j)})$$
$$\leq \prod_{l=0}^{t(j)-1} m(x_{v(j)-t(j)+l}, x_{v(j)-t(j)+l+1})$$
$$\leq \prod_{l=0}^{k-1} m(x_{v(j)-t(j)+l}, x_{v(j)-t(j)+l+1}) \to 1$$

Since $\lim_{j\to\infty} m(x_{n(j)}, x_{v(j)}) = \epsilon$, we deduce from the above that $\lim_{j\to\infty} m(x_{v(j)-t(j)}, x_{n(j)}) = \epsilon$. Now since $\lim_{n\to\infty} m(x_n, x_{n+1}) = 1$ we also have $\lim_{j\to\infty} m(x_{v(j)-t(j)+1}, x_{v(j)-t(j)}) = \epsilon$ and $\lim_{j\to\infty} m(x_{n(j)+1}, x_{n(j)}) = \epsilon$. Now observe by the multiplicative triangle inequality, we also have

$$\frac{\max\{m(x_{v(j)-t(j)}, x_{n(j)+1})^2, m(x_{v(j)-t(j)}, x_{n(j)})^2\}}{m(x_{v(j)-t(j)}, x_{n(j)+1}) \cdot m(x_{v(j)-t(j)}, x_{n(j)})} \leq m(x_{n(j)}, x_{n(j)+1})$$

Now since $\lim_{j\to\infty} m(x_{n(j)+1}, x_{n(j)}) = \epsilon$, and $\lim_{j\to\infty} m(x_{v(j)-t(j)}, x_{n(j)}) = \epsilon$, we deduce from the above that $\lim_{j\to\infty} m(x_{v(j)-t(j)}, x_{n(j)+1}) = \epsilon$. Now observe that

$$\frac{\max\{m(x_{n(j)}, x_{v(j)-t(j)+1})^2, m(x_{v(j)-t(j)}, x_{n(j)})^2\}}{m(x_{n(j)}, x_{v(j)-t(j)+1}) \cdot m(x_{v(j)-t(j)}, x_{n(j)})} \leq m(x_{v(j)-t(j)}, x_{v(j)-t(j)+1})$$

Since $\lim_{n\to\infty} m(x_n, x_{n+1}) = 1$ and $\lim_{j\to\infty} m(x_{v(j)-t(j)}, x_{n(j)}) = \epsilon$, if we take limits in the above we deduce that $\lim_{j\to\infty} m(x_{n(j)}, x_{v(j)-t(j)+1}) = \epsilon$, and similarly, $\lim_{j\to\infty} m(x_{n(j)+1}, x_{v(j)-t(j)+1}) = \epsilon$. Consequently, we deduce that

$$\lim_{j\to\infty} Q(x_{v(j)-t(j)}, x_{n(j)}) = \max\{\epsilon, 1, 1, \sqrt{\epsilon^2}\} = \epsilon$$

$$\lim_{j\to\infty} W(x_{v(j)-t(j)}, x_{n(j)}) = \min\{1, 1, \epsilon, \epsilon\} = 1$$

If we combine the two limiting conditions immediately above with $\lim_{j\to\infty} m(x_{n(j)+1}, x_{v(j)-t(j)+1}) = \epsilon$, then by the continuity of ψ and lower semi-continuity of φ, if we take limits in the inequality below

$$\psi(m(x_{v(j)-t(j)+1}, x_{n(j)+1})) = \psi(m(Tx_{v(j)-t(j)}, Tx_{n(j)}))$$
$$\leq F\left[\frac{\psi(Q(x_{v(j)-t(j)}, x_{n(j)})) \cdot W(x_{v(j)-t(j)}, x_{n(j)})}{\varphi(Q(x_{v(j)-t(j)}, x_{n(j)}))}\right]$$

> **Proof of Theorem C.1 continued 1**
>
> where
>
> $$W(x_{v(j)-t(j)}, x_{n(j)}) = \min\{m(x_{v(j)-t(j)}, x_{v(j)-t(j)+1}), m(x_{n(j)}, x_{n(j)+1}), m(x_{v(j)-t(j)}, x_{n(j)+1}),$$
> $$m(x_{n(j)}, x_{v(j)-t(j)+1})\}^L$$
>
> we deduce that $\psi(\epsilon) \leq F\left[\frac{\psi(\epsilon)}{\varphi(\epsilon)}\right]$ which implies that $\varphi(\epsilon) = 1$, that is, $\epsilon = 1$, a contradiction with $\epsilon > 1$. It follows that $\{x_n\}$ is a multiplicative Cauchy sequence in (X, m). Since (X, m) is complete there exists, there exists $x^* \in X$ such that $\lim_{n \to \infty} x^* = x$. We will prove that $x^* \in \bigcap_{i=1}^{k} A_i$. Since $Y = \bigcup_{i=1}^{k} A_i$ is a cyclic representation of Y with respect to T and $x_0 \in A_1$, we have $\{x_{nk}\}_{n \geq 0} \subseteq A_1$. Since A_1 is closed from $\lim_{n \to \infty} x_n = x^*$, we get that $x^* \in A_1$. Again since $Y = \bigcup_{i=1}^{k} A_i$ is a cyclic representation of Y with respect to T we have $\{x_{nk+1}\}_{n \geq 0} \subseteq A_2$. Since A_2 is closed $\lim_{n \to \infty} x_n = x^*$, we get that $x^* \in A_2$. Continuing this process we have $x^* \in \bigcap_{i=1}^{k} A_i$. Now we will prove that x^* is a fixed point of T. Indeed from $x^* \in \bigcap_{i=1}^{k} A_i$, since for all n, there exists $i(n) \in \{1, 2, \cdots, k\}$ such that $x_n \in A_{i(n)}$. From the multiplicative contractive condition of the theorem, we deduce that
>
> $$\psi(m(Tx^*, x_{n+1})) = \psi(m(Tx^*, Tx_n))$$
> $$\leq F\Big[\psi(Q(x^*, x_n)) \cdot \min\{m(x^*, Tx^*), m(x_n, x_{n+1}), m(x^*, x_{n+1}),$$
> $$m(x_n, Tx^*)\}^L,$$
> $$\varphi(Q(x^*, x_n))\Big]$$
>
> where $Q(x^*, x_n) = \max\{m(x^*, x_n), m(x^*, Tx^*), m(x_n, x_{n+1}), \sqrt{m(x^*, x_{n+1}) \cdot m(x_n, Tx^*)}\}$. Since $x_n \to x^*$, we deduce that
>
> $$\lim_{n \to \infty} Q(x^*, x_n) = m(x^*, Tx^*)$$
>
> and
>
> $$\lim_{n \to \infty} \min\{m(x^*, Tx^*), m(x_n, x_{n+1}), m(x^*, x_{n+1}), m(x_n, Tx^*)\} = 1$$
>
> From the two limiting conditions immediately above, the continuity of ψ and the lower semi-continuity of φ, if we take limits in the inequality below
>
> $$\psi(m(Tx^*, x_{n+1})) = \psi(m(Tx^*, Tx_n))$$
> $$\leq F\Big[\psi(Q(x^*, x_n)) \cdot \min\{m(x^*, Tx^*), m(x_n, x_{n+1}), m(x^*, x_{n+1}),$$
> $$m(x_n, Tx^*)\}^L,$$
> $$\varphi(Q(x^*, x_n))\Big]$$
>
> we deduce that $\psi(m(x^*, Tx^*)) \leq F\left[\frac{\psi(m(x^*, Tx^*))}{\varphi(m(x^*, Tx^*))}\right]$ which implies that $m(x^*, Tx^*) = 1$, that is, x^* is a fixed point of T. For uniqueness of the fixed point, let y^* be another fixed point of T, that is, $Ty^* = y^*$. Now since $Y = \bigcup_{i=1}^{k} A_i$ is a cyclic representation of Y with respect to T, we have $y^* \in \bigcap_{i=1}^{k} A_i$. Now by the multiplicative contractive condition of the theorem, we have, $\psi(m(x^*, y^*)) \leq F\left[\frac{\psi(Q(x^*, y^*))}{\varphi(Q(x^*, y^*))}\right]$. Since x^* and y^* are fixed points of T, we deduce that $Q(x^*, y^*) = m(x^*, y^*)$. Thus it follows that $\psi(m(x^*, y^*)) \leq F\left[\frac{\psi(m(x^*, y^*))}{\varphi(m(x^*, y^*))}\right]$ which implies that $m(x^*, y^*) = 1$, that is, $x^* = y^*$, and the proof is finished.

If we take $p = 1$ and $A_1 = X$ in the previous theorem, then we get the following

Corollary C.2 1

Let (X, m) be a multiplicative complete multiplicative metric space and $T : X \mapsto X$ satisfy the following conditions, there exists $L \geq 0$, $\psi \in \Psi$, $\varphi \in \Phi$ and a multiplicative C-class function F such that

$$\psi(m(Tx, Ty)) \leq F\left[\psi(Q(x,y)) \cdot \min\{m(x,Tx), m(y,Ty), m(x,Ty), m(y,Ty)\}^L, \varphi(Q(x,y))\right]$$

for all $x, y \in X$. Then T has a unique fixed point.

If in the previous theorem, we let the multiplicative C-class function be given by $F(x,y) = \frac{x}{y}$, $\psi(t) = t$, $\varphi(t) = t^{1-k}$, where $k \in (0,1)$, then we obtain the following

Corollary C.3 1

Let $\{A_i\}_{i=1}^{k}$ be nonempty closed subsets of a multiplicative complete multiplicative metric space (X, m) and suppose $T : \bigcup_{i=1}^{k} A_i \mapsto \bigcup_{i=1}^{k} A_i$ satisfies the following conditions (where $A_{p+1} = A_1$)

(a) $T(A_i) \subseteq A_{i+1}$ for $1 \leq i \leq k$

(b) there exists a constant $k \in (0,1)$ such that

$$m(Tx, Ty) \leq Q(x,y)^k \cdot \min\{m(x,Tx), m(y,Ty), m(x,Ty), m(y,Ty)\}^L$$

for all $x \in A_i$, $y \in A_{i+1}$, for $1 \leq i \leq k$. Then T has a unique fixed point that belongs to $\bigcap_{i=1}^{k} A_i$.

3.4 Exercises

Exercise C.1 1

Prove the following: Let $\{A_i\}_{i=1}^{p}$ be nonempty closed subsets of a complete multiplicative metric space (X, m) and suppose $T : \bigcup_{i=1}^{p} A_i \mapsto \bigcup_{i=1}^{p} A_i$ satisfies the following conditions (where $A_{p+1} = A_1$)

(a) $T(A_i) \subseteq A_{i+1}$ for $1 \leq i \leq p$

(b) there exists a constant $k \in (0,1)$ such that $m(Tx, Ty) \leq m(x,y)^k$ for all $x \in A_i$ and $y \in A_{i+1}$, for $1 \leq i \leq p$

Then T has a unique fixed point in $\bigcap_{i=1}^{p} A_i$.

Exercise C.2 1

Prove the following: Let $\{A_i\}_{i=1}^{p}$ be nonempty closed subsets of a complete multiplicative metric space (X, m) and suppose $T : \bigcup_{i=1}^{p} A_i \mapsto \bigcup_{i=1}^{p} A_i$ satisfies the following conditions (where $A_{p+1} = A_1$)

(a) $T(A_i) \subseteq A_{i+1}$ for $1 \leq i \leq p$

(b) there exists a constant $k \in (0,1)$ such that $m(Tx, Ty) \leq (m(x,Ty) \cdot m(y,Tx))^{\frac{k}{2}}$ for all $x \in A_i$ and $y \in A_{i+1}$, for $1 \leq i \leq p$

Then T has a unique fixed point in $\bigcap_{i=1}^{p} A_i$.

Exercise C.3 1

Prove the following: Let $\{A_i\}_{i=1}^{p}$ be nonempty closed subsets of a complete multiplicative metric space (X, m) and suppose $T : \bigcup_{i=1}^{p} A_i \mapsto \bigcup_{i=1}^{p} A_i$ satisfies the following conditions (where $A_{p+1} = A_1$)

(a) $T(A_i) \subseteq A_{i+1}$ for $1 \leq i \leq p$

(b) there exists a constant $k \in (0, 1)$ such that $m(Tx, Ty) \leq [\max\{m(x, Tx), m(y, Ty)\}]^k$ for all $x \in A_i$ and $y \in A_{i+1}$, for $1 \leq i \leq p$

Then T has a unique fixed point in $\bigcap_{i=1}^{p} A_i$

Exercise C.4 1

Prove the following: Let $\{A_i\}_{i=1}^{p}$ be nonempty closed subsets of a complete multiplicative metric space (X, m) and suppose $T : \bigcup_{i=1}^{p} A_i \mapsto \bigcup_{i=1}^{p} A_i$ satisfies the following conditions (where $A_{p+1} = A_1$)

(a) $T(A_i) \subseteq A_{i+1}$ for $1 \leq i \leq p$

(b) there exists $a_1, a_2, a_3, a_4 \geq 0$ with $\sum_{i=1}^{4} a_i < 1$ such that

$$m(Tx, Ty) \leq m(x, y)^{a_1} \cdot m(x, Tx)^{a_2} \cdot m(y, Ty)^{a_3} \cdot [\sqrt{m(x, Ty) \cdot m(y, Tx)}]^{a_4}$$

for all $x \in A_i$ and $y \in A_{i+1}$, for $1 \leq i \leq p$

Then T has a unique fixed point in $\bigcap_{i=1}^{p} A_i$

Exercise C.5 1

Deduce the following

(a) If $p = 1$ and $A_1 = X$ in Exercise C.1, we get the Banach contraction principle [S. Banach, "Sur les operations dans les ensembles abstraits et leur application aux equations itegrales," Fundamenta Mathematicae, vol. 3, pp. 133–181, 1922] in multiplicative metric space

(b) If $p = 1$ and $A_1 = X$ in Exercise C.2, we get the Kannan's fixed point theorem [R. Kannan, "Some results on fixed points," Bulletin of the Calcutta Mathematical Society, vol. 60, pp. 71–76, 1968] in multiplicative metric space

(c) If $p = 1$ and $A_1 = X$ in Exercise C.3, we get the Bianchini's fixed point theorem [R. M. Tiberio Bianchini, "Su un problema di S. Reich riguardante la teoria dei punti fissi," vol. 5, pp.103–108, 1972.] in multiplicative metric space

(d) If $p = 1$ and $A_1 = X$ in Exercise C.4, we get the Hardy and Rogers fixed point theorem [G. E. Hardy and T. D. Rogers, "A generalization of a fixed point theorem of Reich," Canadian Mathematical Bulletin, vol. 16, pp. 201–206, 1973] in multiplicative metric space

Exercise C.6 1

Multiplicative integral was introduced in [A. E. Bashirov, E. M. Kurpinar and A. Ozyapici, Multiplicative calculus and its applications, J. Math. Anal. Appl. 337(1) (2008), 36-48] and we gave an equivalent interpretation [Clement Ampadu, Arslan Hojat Ansari and Memudu Olaposi Olatinwo, FIXED POINT THEOREMS USING MULTIPLICATIVE CONTRACTIVE DEFINITIONS WITH APPLICATION TO MULTIPLICATIVE ANALOGUE OF C-CLASS FUNCTIONS, JP Journal of Fixed Point Theory and Applications, Volume 12, Number 1, 2017, Pages 1-35]. Now letting Ξ be the class of all functions $\xi : [1, \infty) \mapsto [1, \infty)$ such that ξ is a Lebesgue integrable mapping which is summable and $\xi \geq 1$ satisfies $\int_1^\epsilon \xi(t)dt > 1$ for each $\epsilon > 1$, prove that we have the following integral type representation of Theorem C.1

Let $\{A_i\}_{i=1}^p$ be nonempty closed subsets of a complete multiplicative metric space (X, m) and suppose $T : \bigcup_{i=1}^p A_i \mapsto \bigcup_{i=1}^p A_i$ satisfies the following conditions (where $A_{p+1} = A_1$)

(a) $T(A_i) \subseteq A_{i+1}$ for $1 \leq i \leq p$

(b) there exists $L \geq 0$, $\alpha, \beta \in \Xi$, and a multiplicative C-class function F such that

$$\int_1^{m(Tx,Ty)} \alpha(s)ds \leq F\bigg[\int_1^{Q(x,y)} \alpha(s)ds \cdot \min\{m(x,Tx), m(y,Ty), m(x,Ty), m(y,Ty)\}^L,$$
$$\int_1^{Q(x,y)} \beta(s)ds\bigg]$$

for all $x \in A_i$, $y \in A_{i+1}$ for $1 \leq i \leq p$, then T has a unique fixed point $\bigcap_{i=1}^p A_i$

Exercise C.7 1

Taking inspiration from [Mohamed Jleli, Erdal Karapınar, and Bessem Samet, Fixed Point Results for Almost Generalized Cyclic (ψ, φ)-Weak Contractive Type Mappings with Applications, Abstract and Applied Analysis Volume 2012, Article ID 917831, 17 pages]. illustrate Theorem C.1 with an example

3.5 References

(1) Clement Ampadu and Arslan Hojat Ansari, FIXED POINT THEOREMS IN COMPLETE MULTIPLICATIVE METRIC SPACES WITH APPLICATION TO MULTIPLICATIVE ANALOGUE OF C-CLASS FUNCTIONS, JP Journal of Fixed Point Theory and Applications, August 2016, Volume 11, Issue 2, Pages 113-124

(2) W. A. Kirk, P. S. Srinivasan, and P. Veeramani, "Fixed points for mappings satisfying cyclical contractive conditions," Fixed Point Theory, vol. 4, no. 1, pp. 79–89, 2003

(3) S. Banach,"Sur les operations dans les ensembles abstraits et leur application aux equations itegrales," Fundamenta Mathematicae, vol. 3, pp. 133–181, 1922

(4) R. Kannan, "Some results on fixed points," Bulletin of the Calcutta Mathematical Society, vol. 60, pp. 71–76, 1968

(5) R. M. Tiberio Bianchini, "Su un problema di S. Reich riguardante la teoria dei punti fissi," vol. 5, pp.103–108, 1972

(6) G. E. Hardy and T. D. Rogers, "A generalization of a fixed point theorem of Reich," Canadian Mathematical Bulletin, vol. 16, pp. 201–206, 1973

(7) A. E. Bashirov, E. M. Kurpinar and A. Ozyapici, Multiplicative calculus and its applications, J. Math. Anal. Appl. 337(1) (2008), 36-48

(8) Clement Ampadu, Arslan Hojat Ansari and Memudu Olaposi Olatinwo, FIXED POINT THEOREMS USING MULTIPLICATIVE CONTRACTIVE DEFINITIONS WITH APPLICATION TO MULTIPLICATIVE ANALOGUE OF C-CLASS FUNCTIONS, JP Journal of Fixed Point Theory and Applications, Volume 12, Number 1, 2017, Pages 1-35

(9) Mohamed Jleli, Erdal Karapınar, and Bessem Samet, Fixed Point Results for Almost Generalized Cyclic (ψ, φ)-Weak Contractive Type Mappings with Applications, Abstract and Applied Analysis Volume 2012, Article ID 917831, 17 pages

Chapter 4

Fixed Point Theorems for Implicit Cyclic Weak φ_G-Multiplicative Contraction Mappings

4.1 Brief Summary

Abstract D.1 1

We define the multiplicative version of cyclic weak φ_G-contraction mappings implicitly via the C-class function of Ampadu and Ansari [Clement Ampadu and Arslan Hojat Ansari, FIXED POINT THEOREMS IN COMPLETE MULTIPLICATIVE METRIC SPACES WITH APPLICATION TO MULTIPLICATIVE ANALOGUE OF C-CLASS FUNCTIONS, JP Journal of Fixed Point Theory and Applications, August 2016, Volume 11, Issue 2, Pages 113-124]. Consequently, we obtain some fixed point results for such mappings.

4.2 Preliminaries

Definition D.1 1

[M. Pacurar, I. A. Rus, Fixed point theory for cyclic φ-contraction, Nonlinear Analysis (TMA) 72 (2010) 1181–1187] Let X be a nonempty set, k a positive integer, and $T : X \mapsto X$ be an operator. By definition, $\bigcup_{i=1}^{k} X_i$ is a cyclic representation on X with respect to T if

(a) X_i, $i = 1, 2, \cdots, k$ are nonempty sets

(b) $T(X_1) \subseteq X_2, T(X_2) \subseteq X_3, \cdots, T(X_{k-1}) \subseteq X_k, T(X_k) \subseteq A_1$

Notation D.2 1

Ω will denote the class of all continuous functions $G : [1, \infty) \mapsto [1, \infty)$ with $G^{-1}(1) = \{1\}$ such that $G(t_n) \to 1$ implies $t_n \to 1$

Notation D.3 1

Ψ will denote the class of all nondecreasing mappings $\psi : [1, \infty) \mapsto [1, \infty)$ with $\psi^{-1}(1) = \{1\}$ and $\psi(t) < t$ for all $t > 1$

Notation D.4 1

Φ will denote the class of all mappings $\varphi : [1, \infty) \mapsto [1, \infty)$ with $\varphi^{-1}(1) = \{1\}$ and $\varphi(t) < t$ for all $t > 1$ such that $\varphi(t_n) \to 1$ implies $t_n \to 1$

Remark D.5 1

$\Psi \subset \Phi$

Remark D.6 1

Every lower semi-continuous function $\varphi : [1, \infty) \mapsto [1, \infty)$ with $\varphi^{-1}(1) = \{1\}$ and $\varphi(t) < t$ for all $t > 1$ and $\liminf_{t \to \infty} \varphi(t) > 1$ belongs to Ψ

Definition D.7 1

Let (X, m) be a multiplicative metric space, k a positive integer, $A_1, A_2, \cdots, A_k$ be closed nonempty subsets of X and $Y = \bigcup_{i=1}^{k} A_i$. We will say $T : Y \mapsto Y$ is an implicit cyclic weak φ_G-multiplicative contraction if

(a) $\bigcup_{i=1}^{k} A_i$ is a cyclic representation of Y with respect to T

(b) there exists two mappings $\varphi, G : [1, \infty) \mapsto [1, \infty)$ with $G^{-1}(1) = \varphi^{-1}(1) = \{1\}$ and $\varphi(t) < t$ for all $t > 1$ such that

$$G(m(Tx, Ty)) \leq F[G(m(x, y)), \varphi(G(m(x, y)))]$$

for any $x \in A_i$, $y \in A_{i+1}$, $i = 1, 2, \cdots, k$, where $A_{k+1} = A_1$, and $F(x, y) := F(\frac{x}{y})$ is a multiplicative C-class function [Clement Ampadu and Arslan Hojat Ansari, FIXED POINT THEOREMS IN COMPLETE MULTIPLICATIVE METRIC SPACES WITH APPLICATION TO MULTIPLICATIVE ANALOGUE OF C-CLASS FUNCTIONS, JP Journal of Fixed Point Theory and Applications, August 2016, Volume 11, Issue 2, Pages 113-124].

4.3 Main Results

Lemma D.1 1

Let $\varphi \in \Phi$. Then for every closed interval $[a, b] \subset (1, \infty)$, there exists $\alpha \in (0, 1)$ such that $\frac{t}{\varphi(t)} \leq t^{\alpha}$

Proof of Lemma D.1 1

Suppose that for every $\alpha \in (0, 1)$ there exists $t \in [a, b]$ such that $\frac{t}{\varphi(t)} > t^{\alpha}$. Hence for a sequence $\{\alpha_n\}_{n=1}^{\infty} \subset (0, 1)$ with $\lim_{n \to \infty} \alpha_n = 1$, there exists a sequence $\{t_n\}_{n=1}^{\infty} \subset [a, b]$ such that $\frac{t_n}{\varphi(t_n)} > t_n^{\alpha_n}$ for all $n \in \mathbb{N}$. Therefore, $1 \leq \varphi(t_n) < t_n^{1-\alpha_n}$, for all $n \in \mathbb{N}$. Since $\lim_{n \to \infty} \alpha_n = 1$ and $\{t_n\}_{n=1}^{\infty} \subset [a, b]$, then $\lim_{n \to \infty} t_n^{1-\alpha_n} = 1$. It follows that $\lim_{n \to \infty} \varphi(t_n) = 1$. Since $\varphi \in \Phi$, then, $\lim_{n \to \infty} t_n = 1$, and this is a contradiction.

Theorem D.2 1

Let (X, m) be a complete multiplicative metric space, $k \in \mathbb{N}$, $A_1, A_2, \cdots, A_k$ be closed nonempty subsets of X and $Y = \bigcup_{i=1}^{k} A_i$. Suppose that $\varphi : [1, \infty) \mapsto [1, \infty)$ with $\varphi^{-1}(1) = \{1\}$ is a function satisfying " for every interval $[a, b] \subset (1, \infty)$, there exists $\alpha \in (0, 1)$ such that $\frac{t}{\varphi(t)} \leq t^{\alpha}$", and $G \in \Omega$. Let $T : Y \mapsto Y$ be an implicit cyclic weak φ_G-multiplicative contraction mapping. Then T has a unique fixed point $x \in \bigcap_{i=1}^{k} A_i$

Proof of Theorem D.2 1

Let $x_1 \in Y$, and set $x_{n+1} = Tx_n$ for all $n \in \mathbb{N}$. We may assume that $x_1 \in A_1$. Notice that for any n, there exists $i_n \in \{1, 2, \cdots, k\}$ such that $x_n \in A_{i_n}$ and $x_{n+1} \in A_{i_n+1}$. It follows that $x_1 \in A_1, x_2 \in A_2, \cdots, x_k \in A_k, x_{k+1} \in A_1, x_{k+2} \in A_2, \cdots, x_{2k} \in A_k, x_{2k+1} \in A_1, \cdots$. At first we show that $\lim_{n\to\infty} m(x_n, x_{n+1}) = 1$. Observe that, $G(m(x_{n+2}, x_{n+1})) \leq F\Big[G(m(x_{n+1}, x_n)), \varphi(G(m(x_{n+1}, x_n)))\Big] < G(m(x_{n+1}, x_n))$. It follows that the sequence $\{G(m(x_{n+1}, x_n))\}$ is monotone non-increasing and bounded below. Hence, there exists $r \geq 1$ such that $\lim_{n\to\infty} G(m(x_{n+1}, x_n)) = r$. We claim that $r = 1$. If not, suppose $r > 1$, then there exists $\epsilon > 1$ such that $\frac{r}{\epsilon} > 1$. Now observe there exists $N_0 \in \mathbb{N}$ such that for all $n \geq N_0$, $G(m(x_{n+1}, x_n)) \in [\frac{r}{\epsilon}, r\epsilon]$. Since $\varphi: [1, \infty) \mapsto [1, \infty)$ with $\varphi^{-1}(1) = \{1\}$ is a function satisfying " for every interval $[a, b] \subset (1, \infty)$, there exists $\alpha \in (0, 1)$ such that $\frac{t}{\varphi(t)} \leq t^\alpha$", it follows that there exists $\alpha \in (0, 1)$ such that $\frac{t}{\varphi(t)} \leq t^\alpha$ for all $t \in [\frac{r}{\epsilon}, r\epsilon]$. Hence for all $n \geq N_0$, we deduce that $G(m(x_{n+2}, x_{n+1})) \leq G(m(x_{n+1}, x_n))^\alpha$. Since $G \in \Omega$, then taking limits as $n \to \infty$ in $G(m(x_{n+2}, x_{n+1})) \leq G(m(x_{n+1}, x_n))^\alpha$ we get that $G(r) \leq G(r)^\alpha$. Since $\alpha \in (0, 1)$, then $G(r) = 1$ and hence $r = 1$, that is, $\lim_{n\to\infty} G(m(x_{n+1}, x_n)) = 1$, and since $G \in \Omega$, we have $\lim_{n\to\infty} m(x_{n+1}, x_n) = 1$. By the multiplicative triangle inequality, and since $\lim_{n\to\infty} m(x_{n+1}, x_n) = 1$, we deduce that $\lim_{n\to\infty} m(x_{n+l}, x_n) = 1$ for all $l \in \{1, 2, \cdots, k\}$. Now we show that $\{x_n\}$ is a multiplicative Cauchy sequence. Suppose $\{x_n\}$ is not a multiplicative Cauchy sequence, then there exists $a > 1$ and a sequence $\{n(j)\}$ such that $m(x_{n(j+1)}, x_{n(j)}) > a$. Obviously, $n(j) \geq j$ for all $j \in \mathbb{N}$. Since $\lim_{n\to\infty} m(x_{n+l}, x_n) = 1$, there exists $N_0 \in \mathbb{N}$ such that for all $j \geq N_0$, $m(x_{j+1}, x_j) > a^{\frac{1}{3}}$. So for all $j \geq N_0$, $\frac{n(j+1)}{n(j)} \geq 2$, we have

$$a < m(x_{n(j+1)}, x_{n(j)})$$
$$\leq m(x_{n(j+1)}, x_{n(j+1)-1}) \cdot m(x_{n(j+1)-1}, x_{n(j)})$$
$$\leq m(x_{n(j+1)}, x_{n(j+1)-1}) \cdot a$$

If we take limits as $j \to \infty$ in the above, we deduce that $\lim_{j\to\infty} m(x_{n(j+1)}, x_{n(j)}) = a$. Suppose that $t(1) = n(1)$, $t(2) = n(2) + l_2$, where $l_2 \in \{0, 1, 2, \cdots, k-1\}$ such that $t(2) - t(1) \equiv_* 1 \pmod{k}$, $t(3) = n(3)$, $t(4) = n(4) + l_4$, where $l_4 \in \{0, 1, 2, \cdots, k-1\}$ such that $t(4) - t(3) \equiv_* 1 \pmod{k}$, $\cdots$, $t(2j-1) = n(2j-1)$, $t(2j) = n(2j) + l_{2j}$, where $l_{2j} \in \{0, 1, 2, \cdots, k-1\}$ such that $t(2j) - t(2j-1) \equiv_* 1 \pmod{k}$. Observe for all $j \in \mathbb{N}$, we have,

$$a \leq m(x_{n(2j)}, x_{n(2j-1)})$$
$$\leq m(x_{n(2j)}, x_{n(2j)+l_{2j}}) \cdot m(x_{n(2j)+l_{2j}}, x_{n(2j-1)})$$
$$= m(x_{n(2j)}, x_{n(2j)+l_{2j}}) \cdot m(x_{t(2j)}, x_{t(2j-1)})$$
$$\leq m(x_{n(2j)}, x_{n(2j)+l_{2j}})^2 \cdot m(x_{n(2j)}, x_{n(2j-1)})$$

By $\lim_{n\to\infty} m(x_{n+l}, x_n) = 1$ for all $l \in \{1, 2, \cdots, k\}$ and $\lim_{j\to\infty} m(x_{n(j+1)}, x_{n(j)}) = a$, if we take limits in the above inequality we deduce that $\lim_{j\to\infty} m(x_{t(2j)}, x_{t(2j-1)}) = a$, and since $G \in \Omega$, we have $\lim_{j\to\infty} G(m(x_{t(2j)}, x_{t(2j-1)})) = G(a)$. Now observe that

$$m(x_{t(2j)}, x_{t(2j-1)}) \leq m(x_{t(2j)}, x_{t(2j)-1}) \cdot m(x_{t(2j)-1}, x_{t(2j-1)-1}) \cdot m(x_{t(2j-1)-1}, x_{t(2j-1)})$$
$$\leq m(x_{t(2j)}, x_{t(2j)-1})^2 \cdot m(x_{t(2j)}, x_{t(2j-1)}) \cdot m(x_{t(2j-1)-1}, x_{t(2j-1)})^2$$

Since $\lim_{n\to\infty} m(x_{n+1}, x_n) = 1$ and $\lim_{j\to\infty} m(x_{t(2j)}, x_{t(2j-1)}) = a$, if we take limits in the above inequality we deduce that $\lim_{j\to\infty} m(x_{t(2j)-1}, x_{t(2j-1)-1}) = a$, and thus, $\lim_{j\to\infty} G(m(x_{t(2j)-1}, x_{t(2j-1)-1})) = G(a)$. Since $\varphi \in \Phi$ and $t(2j) - t(2j-1) \equiv_* 1 \pmod{k}$ for all $j \in \mathbb{N}$, we have,

$$G(m(x_{t(2j)}, x_{t(2j-1)})) \leq F\Big[G(m(x_{t(2j)-1}, x_{t(2j-1)-1})), \varphi(G(m(x_{t(2j)-1}, x_{t(2j-1)-1})))\Big]$$

Proof of Theorem D.2 continued 1

If $G(a) > 1$, then for some $\epsilon > 1$, $\frac{G(a)}{\epsilon} > 1$. Consequently, we deduce that there exists $N_0 \in \mathbb{N}$ such that for all $n \geq N_0$,

$$G(m(x_{t(2j)-1}, x_{t(2j-1)-1})), G(m(x_{t(2j)-1}, x_{t(2j-1)-1})) \in [\frac{G(a)}{\epsilon}, G(a) \cdot \epsilon]$$

Since $\varphi : [1, \infty) \mapsto [1, \infty)$ with $\varphi^{-1}(1) = \{1\}$ is a function satisfying " for every interval $[a, b] \subset (1, \infty)$, there exists $\alpha \in (0, 1)$ such that $\frac{t}{\varphi(t)} \leq t^\alpha$", it follows that there exists $\alpha \in (0, 1)$ such that $\frac{t}{\varphi(t)} \leq t^\alpha$ for all $t \in [\frac{G(a)}{\epsilon}, G(a) \cdot \epsilon]$. Hence for all $j \geq N_0$, we deduce that

$$G(m(x_{t(2j)}, x_{t(2j-1)})) \leq G(m(x_{t(2j)-1}, x_{t(2j-1)-1}))^\alpha$$

If we take limits in the above inequality we deduce that $G(a) \leq G(a)^\alpha$. Since $\alpha \in (0, 1)$, then $G(a) = 1$, and hence $a = 1$, which is a contradiction. It follows that $\{x_n\}$ is multiplicative Cauchy. Since (X, m) is multiplicative complete, there exists $x \in X$ such that $\lim_{n \to \infty} x_n = x$. From $\lim_{n \to \infty} x_{nk+i} = x$, $\{x_{nk+i} : n \in \mathbb{N}\} \subseteq A_i$ and A_i is closed, we conclude that $x \in A_i$ for $i = 1, 2, \cdots, k$. Therefore $x \in \bigcap_{i=1}^k A_i$. Now we show that $x \in \bigcap_{i=1}^k A_i$ is a fixed point of T. Now observe for all $n \in \mathbb{N}$, we have

$$G(m(x_{n+1}, Tx)) = G(m(Tx_n, Tx))$$
$$\leq F\Big[G(m(x_n, x)), \varphi(G(m(x_n, x)))\Big]$$
$$\leq G(m(x_n, x))$$

If we take limits in the above inequality as $n \to \infty$, we deduce that $G(m(x, Tx)) \leq 1$. It follows that $G(m(x, Tx)) = 1$, and hence $m(x, Tx) = 1$, which implies $x = Tx$. So T has fixed point $x \in \bigcap_{i=1}^k A_i$. Uniqueness of the fixed point follows from the multiplicative contractive condition of the theorem.

Theorem D.3 1

Let $T : Y \mapsto Y$ be as in the previous theorem and $G(t) = t$. Then the fixed point problem for T is well-posed, that is, if there exists a sequence $\{y_n\} \in Y$ with $\lim_{n \to \infty} m(y_n, Ty_n) = 1$, then $\lim_{n \to \infty} y_n = x$ ($x \in \bigcap_{i=1}^k A_i$ is a fixed point of T)

Proof of Theorem D.3 1

Since $x \in \bigcap_{i=1}^k A_i$ and $y_n \in Y$, then from the contractive condition of the previous theorem with $G(t) = t$, we deduce that

$$m(y_n, x) \leq m(y_n, Ty_n) \cdot m(Ty_n, Tx) \leq m(y_n, Ty_n) \cdot F[m(y_n, x), \varphi(m(y_n, x))]$$

Consequently, we deduce that $\lim_{n \to \infty} \varphi(m(y_n, x)) = 1$. Since $\varphi \in \Phi$, then $\lim_{n \to \infty} m(y_n, x) = 1$, and the result follows.

If in Theorem D.2 we take the multiplicative C-class function as $F(x, y) = \frac{x}{y}$, $A_1 = A_2 = \cdots = A_k = X$, and $\varphi(t) = \frac{t}{\psi(t)}$, then we obtain the following which is a multiplicative extension of a result contained in [A. Amini-Harandi, Endpoints of set-valued contractions in metric spaces, Nonlinear Analysis (TMA) 72 (2010) 132–134]

Theorem D.4 1

Let (X, m) be a multiplicative complete multiplicative metric space, and let $T : X \mapsto X$ be a mapping satisfying
$$G(m(Tx, Ty)) \leq \psi(G(m(Tx, Ty))$$
for all $x, y \in X$, where $G \in \Psi$ and $\psi : [1, \infty) \mapsto [1, \infty)$ is upper semi-continuous with $\psi(t) < t$ for all $t > 1$ and satisfies $\liminf_{t \to \infty} \frac{t}{\psi(t)} > 1$. Then T has a unique fixed point.

4.4 Exercises

Exercise D.1 1

Let (X, m) be a multiplicative metric space, k a positive integer, $A_1, A_2, \cdots, A_k$ be closed nonempty subsets of X and $Y = \bigcup_{i=1}^{k} A_i$. We will say $T : Y \mapsto Y$ is an implicit cyclic weak φ_G-multiplicative Kannan contraction if

(a) $\bigcup_{i=1}^{k} A_i$ is a cyclic representation of Y with respect to T

(b) there exists two mappings $\varphi, G : [1, \infty) \mapsto [1, \infty)$ with $G^{-1}(1) = \varphi^{-1}(1) = \{1\}$ and $\varphi(t) < t$ for all $t > 1$ such that
$$G(m(Tx, Ty)) \leq F[G(\sqrt{m(x, Tx) \cdot m(y, Ty)}), \varphi(G(\sqrt{m(x, Tx) \cdot m(y, Ty)}))]$$
for any $x \in A_i$, $y \in A_{i+1}$, $i = 1, 2, \cdots, k$, where $A_{k+1} = A_1$, and $F(x, y) := F(\frac{x}{y})$ is a multiplicative C-class function [Clement Ampadu and Arslan Hojat Ansari, FIXED POINT THEOREMS IN COMPLETE MULTIPLICATIVE METRIC SPACES WITH APPLICATION TO MULTIPLICATIVE ANALOGUE OF C-CLASS FUNCTIONS, JP Journal of Fixed Point Theory and Applications, August 2016, Volume 11, Issue 2, Pages 113-124].

(i) Prove the following: Let (X, m) be a complete multiplicative metric space, $k \in \mathbb{N}$, $A_1, A_2, \cdots, A_k$ be closed nonempty subsets of X and $Y = \bigcup_{i=1}^{k} A_i$. Suppose that $\varphi : [1, \infty) \mapsto [1, \infty)$ with $\varphi^{-1}(1) = \{1\}$ is a function satisfying " for every interval $[a, b] \subset (1, \infty)$, there exists $\alpha \in (0, \frac{1}{2})$ such that $\frac{t}{\varphi(t)} \leq t^{\alpha}$", and $G \in \Omega$. Let $T : Y \mapsto Y$ be an implicit cyclic weak φ_G-multiplicative Kannan contraction mapping. Then T has a unique fixed point $x \in \bigcap_{i=1}^{k} A_i$

(ii) Prove the following: Let $T : Y \mapsto Y$ be as in (i) and $G(t) = t$. Then the fixed point problem for T is well-posed, that is, if there exists a sequence $\{y_n\} \in Y$ with $\lim_{n \to \infty} m(y_n, Ty_n) = 1$, then $\lim_{n \to \infty} y_n = x$ ($x \in \bigcap_{i=1}^{k} A_i$ is a fixed point of T)

(iii) State the Theorem arising from (i) if we take the multiplicative C-class function as $F(x, y) = \frac{x}{y}$, $A_1 = A_2 = \cdots = A_k = X$, and $\varphi(t) = \frac{t}{\psi(t)}$

Exercise D.2 1

Let (X, m) be a multiplicative metric space, k a positive integer, $A_1, A_2, \cdots, A_k$ be closed nonempty subsets of X and $Y = \bigcup_{i=1}^{k} A_i$. We will say $T : Y \mapsto Y$ is an implicit cyclic weak φ_G-multiplicative Chatterjea contraction if

(a) $\bigcup_{i=1}^{k} A_i$ is a cyclic representation of Y with respect to T

(b) there exists two mappings $\varphi, G : [1, \infty) \mapsto [1, \infty)$ with $G^{-1}(1) = \varphi^{-1}(1) = \{1\}$ and $\varphi(t) < t$ for all $t > 1$ such that

$$G(m(Tx, Ty)) \leq F[G(\sqrt{m(x, Ty) \cdot m(y, Tx)}), \varphi(G(\sqrt{m(x, Ty) \cdot m(y, Tx)}))]$$

for any $x \in A_i$, $y \in A_{i+1}$, $i = 1, 2, \cdots, k$, where $A_{k+1} = A_1$, and $F(x, y) := F(\frac{x}{y})$ is a multiplicative C-class function [Clement Ampadu and Arslan Hojat Ansari, FIXED POINT THEOREMS IN COMPLETE MULTIPLICATIVE METRIC SPACES WITH APPLICATION TO MULTIPLICATIVE ANALOGUE OF C-CLASS FUNCTIONS, JP Journal of Fixed Point Theory and Applications, August 2016, Volume 11, Issue 2, Pages 113-124].

(i) Prove the following: Let (X, m) be a complete multiplicative metric space, $k \in \mathbb{N}$, $A_1, A_2, \cdots, A_k$ be closed nonempty subsets of X and $Y = \bigcup_{i=1}^{k} A_i$. Suppose that $\varphi : [1, \infty) \mapsto [1, \infty)$ with $\varphi^{-1}(1) = \{1\}$ is a function satisfying " for every interval $[a, b] \subset (1, \infty)$, there exists $\alpha \in \left(0, \frac{1}{2}\right)$ such that $\frac{t}{\varphi(t)} \leq t^{\alpha}$", and $G \in \Omega$. Let $T : Y \mapsto Y$ be an implicit cyclic weak φ_G-multiplicative Chatterjea contraction mapping. Then T has a unique fixed point $x \in \bigcap_{i=1}^{k} A_i$

(ii) Prove the following: Let $T : Y \mapsto Y$ be as in (i) and $G(t) = t$. Then the fixed point problem for T is well-posed, that is, if there exists a sequence $\{y_n\} \in Y$ with $\lim_{n \to \infty} m(y_n, Ty_n) = 1$, then $\lim_{n \to \infty} y_n = x$ ($x \in \bigcap_{i=1}^{k} A_i$ is a fixed point of T)

(iii) State the Theorem arising from (i) if we take the multiplicative C-class function as $F(x, y) = \frac{x}{y}$, $A_1 = A_2 = \cdots = A_k = X$, and $\varphi(t) = \frac{t}{\psi(t)}$

> **Exercise D.3 1**
>
> Let (X, m) be a multiplicative metric space, k a positive integer, $A_1, A_2, \cdots, A_k$ be closed nonempty subsets of X and $Y = \bigcup_{i=1}^{k} A_i$. We will say $T : Y \mapsto Y$ is an implicit cyclic weak φ_G-multiplicative Reich contraction if
>
> (a) $\bigcup_{i=1}^{k} A_i$ is a cyclic representation of Y with respect to T
>
> (b) there exists two mappings $\varphi, G : [1, \infty) \mapsto [1, \infty)$ with $G^{-1}(1) = \varphi^{-1}(1) = \{1\}$ and $\varphi(t) < t$ for all $t > 1$ such that
>
> $$G(m(Tx, Ty)) \leq F[G(\sqrt[3]{m(x, Tx) \cdot m(y, Ty) \cdot m(x, y)}),$$
> $$\varphi(G(\sqrt[3]{m(x, Tx) \cdot m(y, Ty) \cdot m(x, y)}))]$$
>
> for any $x \in A_i$, $y \in A_{i+1}$, $i = 1, 2, \cdots, k$, where $A_{k+1} = A_1$, and $F(x, y) := F(\frac{x}{y})$ is a multiplicative C-class function [Clement Ampadu and Arslan Hojat Ansari, FIXED POINT THEOREMS IN COMPLETE MULTIPLICATIVE METRIC SPACES WITH APPLICATION TO MULTIPLICATIVE ANALOGUE OF C-CLASS FUNCTIONS, JP Journal of Fixed Point Theory and Applications, August 2016, Volume 11, Issue 2, Pages 113-124].
>
> (i) Prove the following: Let (X, m) be a complete multiplicative metric space, $k \in \mathbb{N}$, $A_1, A_2, \cdots, A_k$ be closed nonempty subsets of X and $Y = \bigcup_{i=1}^{k} A_i$. Suppose that $\varphi : [1, \infty) \mapsto [1, \infty)$ with $\varphi^{-1}(1) = \{1\}$ is a function satisfying " for every interval $[a, b] \subset (1, \infty)$, there exists $\alpha \in (0, \frac{1}{3})$ such that $\frac{t}{\varphi(t)} \leq t^\alpha$", and $G \in \Omega$. Let $T : Y \mapsto Y$ be an implicit cyclic weak φ_G-multiplicative Reich contraction mapping. Then T has a unique fixed point $x \in \bigcap_{i=1}^{k} A_i$
>
> (ii) Prove the following: Let $T : Y \mapsto Y$ be as in (i) and $G(t) = t$. Then the fixed point problem for T is well-posed, that is, if there exists a sequence $\{y_n\} \in Y$ with $\lim_{n \to \infty} m(y_n, Ty_n) = 1$, then $\lim_{n \to \infty} y_n = x$ ($x \in \bigcap_{i=1}^{k} A_i$ is a fixed point of T)
>
> (iii) State the Theorem arising from (i) if we take the multiplicative C-class function as $F(x, y) = \frac{x}{y}$, $A_1 = A_2 = \cdots = A_k = X$, and $\varphi(t) = \frac{t}{\psi(t)}$

4.5 References

(1) Clement Ampadu and Arslan Hojat Ansari, FIXED POINT THEOREMS IN COMPLETE MULTIPLICATIVE METRIC SPACES WITH APPLICATION TO MULTIPLICATIVE ANALOGUE OF C-CLASS FUNCTIONS, JP Journal of Fixed Point Theory and Applications, August 2016, Volume 11, Issue 2, Pages 113-124

(2) M. Pacurar, I. A. Rus, Fixed point theory for cyclic φ-contraction, Nonlinear Analysis (TMA) 72 (2010) 1181–1187

(3) A. Amini-Harandi, Endpoints of set-valued contractions in metric spaces, Nonlinear Analysis (TMA) 72 (2010) 132–134